Seabed Minerals Series
Volume 5

Selection of Sites for Seabed Manganese Nodule Processing Plants

Seabed Minerals Series

Volume 1. **Assessment of Manganese Nodule Resources**

Volume 2. **Analysis of Exploration and Mining Technology for Manganese Nodules**

Volume 3. **Analysis of Processing Technology for Manganese Nodules**

Volume 4. **Delineation of Mine Sites and Potential in Different Sea Areas**

Volume 5. **Selection of Sites for Seabed Manganese Nodule Processing Plants**

Volume 6. **Financial Analysis of a Manganese Nodule Mining Venture: A Balance Sheet**

Volume 7. **Economics of Lateritic Nickel Production in Developing Countries**

Volume 8. **Regulatory Framework for Manganese Nodule Development**

Volume 9. **Review of Activities of International Consortia and other Public and Private Entities**

Seabed Minerals Series
Volume 5

Selection of Sites for Seabed Manganese Nodule Processing Plants

United Nations

Published in co-operation with the
United Nations by

Graham & Trotman
A member of the Kluwer Academic Publishers Group
LONDON/DORDRECHT/BOSTON

Graham & Trotman Limited
Sterling House
66 Wilton Road
London
SW1V 1DE
UK

Graham & Trotman Inc.
Kluwer Academic Publishers Group
101 Philip Drive
Assinippi Park
Norwell, MA 02061
USA

in co-operation with the United Nations

ISBN 0 86010 395 1 (Volume)
ISBN 0 86010 927 5 (Series)

First published in 1989

British Library Cataloguing in Publication Data
Selection of sites for seabed manganese nodule processing plants
1. Manganese nodules. Processing
I. United Nations II. Series
622'.34629

Library of Congress Cataloging-in-Publication Data
Selection of sites for seabed manganese.
(Seabed minerals series; v. 5)
Includes bibliographical references.
1. Manganese—Metallurgy. 2. Metallurgical plants—Location. 3. Manganese nodules. I. United Nations.
II. Series.
TN799.M3S45 1989 338.2'74629 89-23405

Typeset in Baskerville by Advanced Filmsetters (Glasgow) Ltd
Printed and bound in Great Britain by Bell & Bain Ltd, Glasgow

Contents

Preface to the Series

The seabed is one of the newest and potentially most rewarding frontiers that has challenged mankind in its quest for knowledge and material achievement. Resources of the deep seabed promise to make an enormous contribution to the world's resource base if their potential is realized. At the present time, the resources of the deep seabed of immediate interest are in the form of manganese nodules which lie on the surface of the ocean floor and contain numerous metals—copper, nickel, cobalt, manganese, molybdenum, vanadium and titanium.

In addition to the potential for increasing the world's resource base, these minerals are particularly intriguing because they lie beyond the limits of national jurisdiction—they belong to no nation. According to a resolution adopted by the General Assembly of the United Nations, the area of the seabed and the resources therein are the common heritage of mankind.

While, for many years, manganese nodules have been the domain of marine geologists, chemists and lawyers, interest is currently being shown by a broader audience—corporate and public entrepreneurs, policy-makers, managers, investors, engineers and technicians, as well as the international community of persons interested in the mineral and ocean industries, strategic minerals, and global development. The past two decades have witnessed the development of technology with a view to mining manganese nodules from a depth of nearly 15 000 feet and to extract the economically important metals from them. It has also witnessed the development, in the Third United Nations Conference on the Law of the Sea, of the legal and institutional framework in which the exploitation of the nodules will take place—who will exploit these resources, how this will be done, and who will benefit from this exploitation. At this moment, a water-shed appears to have been reached when decisions about the application of technological prototypes of commercial operations are being taken and when the negotiations in the Law of the Sea Conference culminated in the adoption of the United Nations Convention on the Law of the Sea. Thus, it seems to be an appropriate time to take stock, to assimilate widely diverse and dispersed information and analyses, and to make preparations for a future which envisages the resources of this newest frontier being utilized for the benefit of mankind as a whole.

It is in this context that the United Nations has undertaken the task of preparing a series of nine volumes examining different aspects of the development of manganese nodule resources.

The series starts with a discussion of how many of these resources exist in the world ocean. The quality and the quantity of the data collected through scientific expeditions and prospecting and exploration cruises are examined, the methods of estimating resources and reserves and their distribution on the basis of these data are explained, and a review of the currently available estimates is presented.

Since new technology had to be developed and tested for exploring, mining and processing manganese nodules, the next two volumes in the series describe and evaluate these technologies in terms of their efficiency and reliability.

Mine development is preceded by the delineation of a mine site. The fourth volume discusses the criteria used to designate a mine site. A potential mine site is defined as satisfying certain economic, geological and technological criteria; the size of the area may vary according to these criteria. This volume also develops and applies a simulation model which calculates the variations in the size of the mine site resulting from the variations in these criteria. Finally, based on available data and with the application of the simulation model, this volume attempts to identify areas meeting given sets of criteria.

The location of a processing plant for manganese nodules depends more on the availability and cost of complementary inputs (e.g., energy and chemicals) than on the traditional considerations of proximity to the mine or to the market. The criteria for selecting a site for a processing plant are examined in the fifth volume and a few likely sites of first generation processing plants are assessed according to these criteria.

The economics of a manganese nodule mining project are the crucial factor in the development of these resources. Currently, the economics can at best be based on order-of-magnitude estimates. These estimates are carefully examined and a financial profile of a potential seabed mining company is constructed in the sixth volume.

The seventh volume attempts to compare the economics of a manganese nodule mining project with alternative mining projects, e.g. nickel laterites.

The geological, technological and economic considerations for resource development can be translated into practice only within a defined regulatory framework. The eighth volume reviews the regulatory framework for the exploration and exploitation of seabed resources that is being devised by the international community. Finally, private companies, joined in several consortia, and public entities have already initiated plans and are carrying out preliminary work that may lead to mine development—the last volume of the series presents a summary of their activities and the status of their work.

The series is prepared by the United Nations Office for Ocean Affairs and the Law of the Sea (the Office in which most activities in marine affairs carried out in the United Nations Secretariat, including those of the former Ocean Economics and Technology Branch of the United Nations Department of International Economic and Social Affairs, have been consolidated) together with leading external experts. It draws and builds upon the information, data and expertise that the United Nations has been developing for over a decade concomitant with its work in the Law of the Sea Conference, and currently, in the Preparatory Commission for the International Sea-bed Authority and for the International Tribunal for the Law of the Sea. These capabilities have been utilized in, and at the same time strengthened by, the development and maintenance at United Nations

Headquarters of a data base on manganese nodules and a literature data base on marine minerals as well as the convening of a Group of Experts Meeting, the proceedings of which were published as *Manganese Nodules: Dimensions and Perspectives* (Reidel, Dordrecht, 1979).

The efforts of the United Nations will succeed if the series can contribute to a wider understanding of seabed resources and the problems and opportunities involved in their development, and can encourage constructive ideas and actions.

Acknowledgements

We gratefully acknowledge the contribution of Dr Herbert Drechsler of HDD Resource Consultants in the preparation of this volume. We wish to thank Dr Jean-Pierre Lenoble of the Institut Français de Recherche pour l'Exploitation de la Mer (IFREMER), Dr C. J. Johnson of the East–West Center and Mr D. W. Pasho of Energy, Mines and Resources, Canada, for reviewing the manuscript and offering many helpful comments and suggestions.

Chapter 1

INTRODUCTION

Although a metal producing facility on the scale of a nodule processing plant has the ability to significantly alter economic growth in a particular locality, capital requirements for such a venture are considerable. They range from a low estimate of US$ 1021.2 million for a three-metal venture to US$ 1775 million for the highest estimate of a four-metal venture (in 1981 US dollars). Capital costs for the processing segment of an integrated nodule venture have been estimated to account for approximately 50% of total capital requirements.

In undertaking an initial assessment of the potential economic impact of establishing a manganese nodule processing plant in Hawaii, the Department of Planning and Economic Development of the State of Hawaii concluded, *inter alia,* that during the construction phase of the plant, the increased output of the county of Hawaii would range from US$ 119 million to US$ 221 million while the increase in the total output of the State would range from US$ 142 million to US$ 264 million (in 1981 US dollars). During the operational phase, the increase in the total output of the county of Hawaii was estimated to range from US$ 535 million to US$ 836 million, while the increase in the total output of the State was estimated to range from US$ 572 million to US$ 880 million (in 1981 US dollars).[1]

The contribution of the mineral sector to civilization, especially in the Industrial Age, and to material progress is immeasureable: all major advances in meeting human needs in the past—for food, housing, health care, education, jobs and transport—have been predicated upon the use of additional minerals, more efficient use of minerals, or use of better minerals—indeed, very often a combination of all three.

From the mid-eighteenth century until World War I, economic history centres around a group of changes known as the Industrial Revolution. At the heart of the Industrial Revolution was a cluster of innovations in the technique and mode of industrial production:

(1) the substitution of inanimate for animal sources of power (in particular, the introduction of steam power fuelled by coal);
(2) the substitution of machines for human skills and strength;
(3) the invention of new methods for transforming matter (in particular, new ways of making iron and steel and industrial chemicals); and
(4) the organization of work in large, centrally powered units (factories, forges, mills) that made possible the immediate supervision of the production processes and a more efficient division of labour.

The first country to make the transition to modern industry and economic growth was Britain. The reasons for this precocity were partly material, partly social and institutional. Favoured by its oceanic position, and by the abundance of those natural resources that proved to be particularly important for the new industrial technology, Britain, with its modest size but highly indented coastline, was the first country to take advantage of the opportunities offered. The natural resources of significance included: water (for water power, chemical processes and textile finishing); salt (for the manufacture of industrial alkalis and acids); china clay, iron and nonferrous ores; and above all, huge accessible deposits of coal for use as fuel and as a source of carbon in the smelting of iron.

Domestic demand in Britain (which at the beginning of the eighteenth century was the biggest single market in the world at that time) reinforced by a large and growing demand for British products on the continent of Europe placed heavy strains on the existing techniques and organization of production. One of the first industries to feel the pressure was mining, where the deeper one dug, the more difficult it became to drain the infiltrating underground water from the pits. Another industry to feel the strain of increased demand was iron-making. Here, the most acute bottleneck was the shortage of inexpensive fuel and carbon.

Following the discovery of coke smelting and with the availability of more powerful bellows, British production of pig iron started to grow rapidly. By the end of the eighteenth century, the combined techniques of puddling and rolling had brought the industrial revolution to iron manufacturing.

The other countries of Europe took longer to effect their industrial revolutions. Belgium was first in the 1820s and 30s, with a marked emphasis on coal mining and metallurgy. France followed and, being poor in coal, concentrated on textiles and other light industries. Finally, Prussia, with large coal resources in Silesia and Westphalia, developed from the 1840s on a pattern similar to that of Belgium.

During the nineteenth century, industrial growth in Western Europe and the United States was based largely on indigenous mineral resources. Europe, for all practical purposes, was self-sufficient in base metal supplies with the United Kingdom, in particular, providing over half of the world's supply of copper.[2] For the industrialized countries of Western Europe, with markets for metals already established and with indigenous mineral resources and the ability to raise capital for their development, the issue of the location of processing facilities or, smelters and refineries, was resolved by siting such plants in the industrialized country that could offer the best incentives to an enterprise that wanted to refine metal for manufacturing.

By the beginning of the twentieth century, the industrial nations[3] were reaching out overseas in all directions to support their growing industrial activities and investing their savings in the development of mineral resources. Many of the areas that were suitable for mineral development were within the colonial empires. Capital was relatively easy to obtain for these areas because the political and economic structures were closely tied to those of the metropolitan centres of the industrialized world. Moreover, the mineral resources in these areas were richer and larger. The copperbelt of central Africa, for example, grew relatively well under conditions which made it almost an extension of the metropolitan European scene.

After the Second World War, the growth of pressures for political sovereignty and the general opposition to colonialism and empire resulted in many of these previously metropolitan-controlled areas obtaining their political independence. This had the effect of weakening the fabric of the economic and financial relationships that had facilitated mineral development in the earlier period. In some parts of the world, strong feeling arose against what was seen as foreign "exploitation" of local resources by large international companies. This contributed to disengagement by the metropolitan countries, in part voluntarily by investors, in part encouraged by emerging countries who showed a strong desire to exercise ever-greater control over the proceeds arising from the exploitation of their resources, and to try to obtain the value added from processing these resources locally.

As part of the effort to translate their desires into real action, existing mines in some developing countries were nationalized and, during the past decade, as the full implications of downstream activities in the mining industry were better understood, proposals were submitted in international meetings to, *inter alia*, increase the levels of local processing of mineral resources within the developing countries.[4] A statement on this issue, by the 1974 Latin American Conference on Industrialization made the following points:

> National industrialization policies should stress the objective of increasing the external autonomy of the developing regions and countries, with special attention to the promotion of exports ... and also seek to increase the value-added to raw materials being processed and exported. In the light of the foregoing, it is proposed that national industrialization policies should: (a) promote integrated industrialization based on the potential of

> each country; (b) stimulate by various means the intensive use of natural resources, both through the promotion of employment policies and through the formulation of schemes which favour the processing of available raw materials.[5]

Among the reasons stated by the developing countries for seeking to increase processing of raw materials are: (a) national strategies for industrialization based on the use of local raw materials; (b) the need to lessen the dependence on the industrial countries; (c) the development of skills that can be used in other sectors of the economy; (d) the ability to enjoy more of the economic rents resulting from raw material production; and (e) the opportunity to obtain investment capital which might not otherwise be available.

Meanwhile, for a quarter of a century after the Second World War, private mineral companies operating in developing countries were evolving their own policies. Partly as a result of the overall demands of the developing countries but also as a result of the challenge to established methods of financing mineral development, the industrialized countries of Western Europe began to shift their mining interests away from the developing world to the United States, Canada, Australia and South Africa. With the growth of Japan as a modern industrial power, short on indigenous mineral resources but with a great demand for metals to supply the global needs of its semi-manufactures and manufactures, the competition for secure sources of minerals has recently led to interest in the deep seabed.[6,7]

Interest in the deep seabed poses a number of challenges to the international community. First, the area under consideration has been declared the common heritage of mankind and the benefits from exploiting the resources of this area are to be shared among all mankind; second, the act of exploiting the mineral resources of this area may threaten the position of some established developing and industrialized land-based producer States; and, third, because the innovative programmes and research activities undertaken by private groups, consortia and State enterprises in industrialized and semi-industrialized countries, to develop the technological capability to exploit the resources of the deep seabed was in part a reaction to the demands from the developing countries to obtain a greater share of the proceeds from the exploitation of their own land-based resources.

Moreover, within the context of the state of technological readiness, developing countries will not be at the forefront of the manganese nodule mining industry.[8] Short of their participation in the activities of the Enterprise (the operating arm of the International Sea-Bed Authority, the organization established by the United Nations Convention on the Law of Sea, through which States Parties shall organize and control the exploration and exploitation of deep seabed resources), which should help in the development of some aspects of their technological capabilities, there appears to be only a limited number of areas that comprise the nascent nodule mining industry in which developing countries can meaningfully participate in its initial stages. One possible area of participation is in the processing of manganese nodules, that is metal refining. One mechanism for ensuring such participation could be through competition for processing plant sites.

Although energy requirements for manganese nodule processing plants will vary depending on the scale of operations (from 1 to 3 million tons per year of dry nodules), the processing route selected (pyrometallurgical or hydrometallurgical) and the number of valuable metals to be extracted from nodules (3 or 4 metals), such plants are most likely to be large, energy intensive operations. Proximity to the source of this consumable as well as to water, to a disposal facility for process wastes and to a technical labour force will all be factors that enhance the desirability of one site over another.

Some of the developing countries, at the present time, enjoy comparative advantages over some of the industrialized countries in respect of lower transport costs from nodule mining sites, less stringent environmental controls and associated costs, energy availability, and lower costs for complementary inputs to a nodule processing operation.

This study addresses itself to an analysis of the factors that would affect the choice of a site for a nodule processing facility. It reviews the mining industry's record with respect to the location of processing facilities in developing countries and reports on efforts made by the latter group of countries in international fora to redress an imbalance in the global distribution of processing facilities. It explains some of the key factors that affect the decision on the location of a plant by an investor and reviews the myriad of possible objectives of developing countries regarding the acquisition of such a plant. The site requirements for a nodule processing plant are discussed from the literature available on the technology for processing manganese nodules into the major metals of interest, and the procedures for determining the location of a land-based processing site are then presented. Finally, the potential locations for first generation operations are identified.

This study has uses beyond the merely locational considerations for a nodule processing facility. The various factors that are reviewed can well serve in the analysis of sites for a processing facility whose source of ore is terrestrial.

NOTES

[1] Hawaii Department of Planning and Economic Development. 1981. *Feasibility and potential impact of manganese nodules processing in the Puna and Kohala districts of Hawaii.* Prepared by the Department of Planning and Economic Development, State of Hawaii and the US Office of Ocean Minerals and Energy, Honolulu: Department of Planning and Economic Development, State of Hawaii.

[2] Copper production from the United Kingdom was chiefly from mines in Cornwall, Devon and Anglesey. These mines were depleted around 1875.

[3] The effect of the localization of processing facilities in the industrialized countries is very marked on the foreign exchange earnings. Mineral exporting developing countries without processing obtain a much lower value per unit of metal contained in their ores. For example, copper wire and cable has a value seven to nine times greater per unit of copper than the copper ore at the mine.

[4] United Nations General Assembly resolution 3362 (S-VII), *Development and International Economic Co-operation,* 16 September 1975; UNCTAD, 1976. *New Directions and New Structures for Trade and Development* (TD/183), pp. 19–23.

[5] United Nations Industrial Development Organization, *Industrialization in Latin America,* Declaration of the Latin American Conference on Industrialization, Mexico City, December 1974.

[6] The concept of security of supply from a national standpoint has led to the development of different yardsticks that supplement the technical and economic analysis of projects.

[7] Although Japan's mineral deposits are fairly diverse, with a few exceptions, the reserves are small and production is inadequate to meet more than a small part of domestic requirements. Coal, copper, zinc, lead and silver are among the most important minerals. There is an almost complete lack of nickel, cobalt, bauxite and oil.

[8] India is an exception. It was the first country to register as a Pioneer Investor under Resolution II of the Third United Nations Conference on the Law of the Sea.

Chapter 2

METALLURGICAL PLANTS IN DEVELOPING COUNTRIES

Development, as conceived of and pursued by developing countries, is a many-faceted concept. Broadly speaking, development goals can be said to consist of: (a) not only growth as such, but also of the establishment of capacities needed to permit the country to initiate and pursue growth policies of its own; (b) distribution of the growth among major population strata; and (c) economic diversification, to help the country to avoid the disadvantages of one-sided dependence. The three major goals, that is, capacity for growth, distribution, and diversification, can be broken down into sub-goals such as an increase in earnings of foreign exchange, an increase in government revenues and in the volume of investment, creation of employment, training of local personnel and national institution building. For mineral-dependent economies, the sub-goals may include a greater emphasis on production and exports of manufactures and semi-manufactures as opposed to crude materials; more diversity in terms of companies responsible for exploiting and treating their mineral resources; and a greater variety of export markets for their mineral products.

Numerous developing countries depend heavily on the production and sales of minerals for their economic development. A large proportion of the minerals produced in the developing countries is exported in crude form, to be processed abroad. Though several factors suggest that mineral processing in the exporting developing countries should be cheaper than elsewhere (some factors may work in the opposite direction however), the major economic arguments that have been proposed to explain the large gap in mineral processing capacities of developing countries include, *inter alia,* the need to put up an expensive infrastructure and the industrial inexperience in these countries.

On the political side, the primary argument used to justify the lack of processing facilities in the developing mineral economies has been that the uncertainties concerning the future actions of the governments of said countries have created an environment that threatens the "availability for use" of the mineral in question to the major consumers. As the argument goes, since the countries in question control the production of reserves of the mineral commodity, they could, if they had the ability to forego export earnings, have an effective means to implement policies in line with their motives. When developed to the extreme, the argument concludes that through cartel action the developing country mineral-economies could deny or limit supplies of a given mineral.

Since similar arguments may be used in the discussions on the merits and demerits of locating manganese nodule processing plants in developing countries, it would appear prudent to review these arguments and the parallels that exist between a land-based processing facility for terrestrial ores and one for manganese nodules.

ECONOMIC ARGUMENTS AGAINST THE LOCATION OF PROCESSING FACILITIES IN DEVELOPING COUNTRIES

Processing minerals in developing countries is said to be more expensive than in the industrialized countries because the required infrastructure has to be established from scratch. In contrast, however, it is said that processing installations in industrialized countries can make use of an already existing infrastructure and, therefore, result in a lower price for refined metal. This argument is probably exaggerated in many cases. First, much of the infrastructure already established in the developing countries to serve the mining operations (that is, railways, roads, harbours, power and water facilities) can also be used by the processing industry. Hence, the requirements for additional infrastructure may be somewhat marginal. Secondly, where the industry is charged the full cost for the infrastructure, including the annualized capital costs based on the replacement value of installations, there should not be any difference in the amounts charged, irrespective of whether the facilities are old or have just been established. That such a difference exists can be ascribed either to accounting procedures that impose an unduly heavy charge on the use of new facilities in developing countries or to an implicit subsidization by the owners of infrastructure in industrialized countries.

A valid argument about the relative cheapness of infrastructure in industrialized countries is that there will be more users of the infrastructure because of such countries' industrial diversification. This point appears to suggest that the higher cost of infrastructure in developing countries is a temporary infant-industry phenomenon, which will disappear with further industrialization.

The cost of processing in developing countries is also said to be more expensive because of the industrial inexperience of such countries. Thus, it is said that there would be greater costs for training the labour force, that more inputs would have to be imported, making the product more expensive in many cases, and finally that the productivity of the expatriate experts brought in to manage the construction and running of the plant would be reduced because of their unfamiliarity with the social and economic environment. There may be truth in this argument, but as in the case of infrastructure costs, one may query its importance. First, though the cost of training and of expatriate management may be high and the productivity per man-unit low, it will at least partly be offset by the low wages in developing countries. Secondly, exactly the same argument should apply to other industrial and mining ventures in the developing countries. Yet the costs of production in many mines in developing countries compare favourably with the cost of exploiting deposits of similar quality in the industrialized world. To the extent that a cost difference on account of industrial inexperience really remains, here too, there is cause to regard it as a temporary, infant-industry phenomenon.

Obviously, the economy of each manganese nodule processing facility must be studied on its own merits. The above discussion suggests, however, that there is little ground for believing that processing costs in the long run would be higher in developing country installations than elsewhere. In fact, the arguments presented here suggest that if the cost of transporting nodules from a mine site to a given harbour is not prohibitive, then, once experience has been gained, other costs associated with processing nodules may well be lower in a developing country. Hence, in an economically unbiased world, investments in future nodule processing facilities in developing countries should in some cases generate at least normal returns on capital with ensuing normal contributions to earnings of foreign exchange and government revenues.

POLICIES THAT INHIBIT THE LOCATION OF PROCESSING FACILITIES IN DEVELOPING COUNTRIES

A good deal of the discussion of mineral processing in developing countries has focussed on "barriers" or obstacles to the development of processing facilities. These barriers are generally perceived to be policy distortions or market imperfections which have resulted in a lesser amount of processing in such countries than would have resulted from the unimpeded operation of competitive market forces. Barriers to processing are often described as either "artificial" or "natural". Artificial barriers include trade-distorting policies introduced by the industrialized countries, restrictive business practices of transnational corporations and production distorting policies of the developing countries themselves. The natural barriers, on the other hand, are said to be the underlying economic characteristics of particular developing countries.

Tariffs and Other Trade Limitations

The tariff structures of the industrialized countries frequently impose higher rates of duty on the importation of processed materials than on unprocessed ore. The situation can, in theory, result in a very high rate of effective protection for processing operations, with the result that further processing in developing countries is discouraged. In practice, however, it is not clear that many developing countries are significantly affected by this tariff escalation, at least in respect of smelting and refining operations. The combined effect of the tariff-reducing negotiations of the General Agreement on Tariffs and Trade (GATT), the Generalized System of Preferences which is applied by most western countries to developing country exports, and the preferential-access provisions of the Lomé Convention in respect of exports to the European Economic Community has reduced tariff barriers on imports of refined metal from developing countries to insignificant levels.

Similarly, while non-tariff barriers to trade, such as the quantitative import restrictions imposed by certain industrialized countries, can, in theory, have a deterrent effect on developing country processing, there is little concrete evidence to show that such non-tariff barriers do in fact have such an effect in the specific case of non-fuel minerals.

Market Distortion by Transnational Corporations

For a number of reasons, processing facilities in developing countries are disproportionately small when compared with their numbers in the industrialized countries. A transnational corporation (TNC) will normally attempt to reduce risk by diversifying its investments and in particular will attempt to avoid a concentration of investment in countries thought to be prone to nationalization. Transnational corporations may also be subject to pressure from their home governments, whose defence and strategic interests or whose concern with maintaining domestic employment levels may favour expansion, or at least maintenance, of their home processing capacities.

These non-economic reasons are in addition to such purely economic considerations as the fact that a TNC may have different costs than a developing country, because the corporation has easier access to world capital and raw materials markets, or that the TNC may face, within its own vertically integrated operations, a set of marginal costs that differ from the world prices or the costs faced by the producing country.

The extent to which these factors limit a mineral-producing country's ability to establish processing facilities depends greatly on the extent of corporate concentration in the particular mineral industry.

TNCs can restrict developing countries' processing potential through the use of export-restraining business practices. Such practices include restrictions on exports from developing-country plants or even complete refusal by the TNC to permit such exports. In the case of most minerals, however, there is little clear evidence that TNCs have used such monopoly power to restrict developing country exports, at least in the past decade.

A more specific use of monopoly power which is sometimes cited as a barrier to developing country processing is the tendency of shipping conferences to increase freight rates for processed products solely because of the ship owner's superior bargaining power *vis-a-vis* the exporting countries, and without any justification based on the increased cost of handling processed material.

A final aspect of monopoly power sometimes cited as a barrier to developing country processing is the use of massive advertising by TNCs to create brand-name loyalty that is not necessarily justified on the basis of product quality differences. In the case of minerals, however, advertising is of relatively little importance, since quality standards for refined metals are normally set by the various materials testing organizations or by the metal exchanges themselves. Once a producer has its brand certified as good for delivery on the relevant metal exchange, little further assurance of basic quality is required. Some customers may prefer deliveries from industrial-country suppliers, either because transportation and delivery are thought to be more reliable or because the buyer prefers a specific brand for particular end-users.

The barriers to further processing of minerals in developing countries cited above are not so much impenetrable roadblocks as they are factors which tend to have a stronger impact on developing-country projects than on those in industrialized countries. Rather than saying that such barriers are either so important that one cannot think about further processing in developing countries, or, on the other hand, that the barriers are of little or no importance, a more fruitful approach may be to consider in detail the economics of nodule processing facilities and the special characteristics of nodule mining that could make such plants viable in developing countries.

SALIENT TRENDS IN THE LOCATION OF METALLURGICAL PLANTS

In an analysis of the degree of processing conducted in different regions of the world, based on nine minerals consisting of tin, copper, lead, nickel, zinc, iron, manganese, phosphorous and bauxite during the years 1950, 1960 and 1970, it was concluded that only about 30% of the minerals mined in the developing countries are processed there (Table 1).[1] In the industrialized countries, particularly in Japan, increases in processing capacity have outstripped increases in mining capacity. In both Australia and South Africa, mining capacity has expanded much more rapidly than processing facilities, with a resultant increase in the quantities of ores and concentrates shipped. Within the centrally planned economies, processing capacity has been maintained commensurate with mining capacity. Noting that this ratio for developing market economies had remained relatively constant over two decades, Bosson and Varon concluded that increases in processing facilities in developing countries were merely to meet increases in mine capacity.

Both developed and developing countries recognize the scope for increasing value added through processing in de-

Table 1. Degree of processing conducted in different regions for major minerals combined, for 1950, 1960, and 1970[a]

Region	*Minerals processed as a percentage of total minerals mined in each region*[b]		
	1950	*1960*	*1970*
United States and Canada	146	179	179
Western Europe	250	250	295
Japan	235	381	1046
Australia and South Africa	89	72	38
Developing Market Economies	30	28	29
Centrally Planned Economies	99	102	108

Source: Bosson, R. and Varon, B. op. cit.

[a] Bauxite and aluminium, copper, iron ore, lead, manganese ore, nickel, phosphate rock, tin and zinc.

[b] The degree of processing is defined as the ratio of the value of mineral produced over the value of mineral mined if processed to its final stage, namely metal ingot. The ratio was computed as follows:

$$\text{Degree of processing} = \frac{A-B}{C-B} \times 100$$

where: A = value produced (mining and processing value);
B = mine value of production; and
C = value produced, had all ore mined been processed to refined metal except for iron ore, manganese ore, and phosphate rock for which pelletized or sinterized iron ore, ferromanganese, and super phosphorite fertilizer, respectively, were taken as representing the processed product.

veloping countries. Moreover, the insistence on processing is not limited to the developing countries, but forms a key component of the current resource policies of both Canada and Australia. The shift to processing in developing countries will depend on the nature of the future partnership between producing and consuming countries, particularly as it affects transfer of technology and capital, control of markets and marketing.

The opportunity offered by the location of a processing plant for manganese nodules in a developing country consists of: (1) securing the training and employment of personnel from that country; (2) selling goods and services originating from that country to the facility; and (3) utilizing the metal products originating from that facility in producing semi-manufactures and manufactures. Since manganese nodule processing plants will be large energy intensive operations, employment objectives directly related to the plant may not be the most rewarding economically. Other possibilities are opened up by the latter two categories of benefits.

NOTE

[1] Bosson, R. and Varon, B. 1977. *The Mining Industry and the Developing Countries*. Oxford University Press for the World Bank.

Chapter 3

PROJECT ECONOMICS

Deep seabed nodule mining will involve unique mining methods that require new technology. Much research and successful testing has been conducted by private, semi-private and public enterprises on small-scale nodule mining and processing systems and, although no full-scale mining operation has yet been undertaken, at the present time, it is generally accepted that the basic technology for nodule mining and processing have been established.[1,2]

Since the late 1970s, various studies, based on the fact that these technologies had been developed and their operating parameters were available, investigated the commercial viability of an integrated nodule venture. The studies indicate that nodule mining and processing will be unlikely to meet the high internal rates of return (also referred to as discounted cash flow rates of return) required for such risky investments.[3-6] Nevertheless, since that time, no major changes have been made in the mining and processing technologies that would fundamentally alter the previous economic analysis.

Previous economic analyses indicate that from an investment and operating cost perspective, nodule mining operations will basically comprise processing operations. Tables 2 and 3 contain a summary of selected seabed mining cost models since 1980. Due to the different assumptions made by the various authors in respect of, *inter alia*, the choice of mining equipment and metallurgical process, location of processing plant and number of mining and transport vessels, estimates of capital requirements differ. Despite this wide range of capital costs, the total capital requirements may be broken down as follows:

processing would account for approximately 50% of total capital investments; mining would account for 30%, while transportation would account for the remainder of total capital costs.

The proportional breakdown of total annual operating costs reflects the same dominance of the processing operations. For example, when the processing route selected is for the recovery of three metals (the three metal option that produces copper, nickel and cobalt), annual direct operating costs of these operations are in the order of 50% of total annual operating costs. In the four metal option, with a ferromanganese plant utilizing manganese to produce ferromanganese, the annual operating costs for the entire processing segment is estimated to range up to 75% of total annual direct operating costs of the venture.

Within the processing segment of the venture, operating costs in respect of utilities and fuel loom large. Estimates of annual operating costs in this regard range from a low of 15% of total annual operating costs in a three metal venture to a high of 34% in a four metal venture.

With regard to the economic return of an integrated venture, the various models show a range in discounted cash flow rates of return of from 4.1 to 9.2% for a three metal option and from a negative rate of return of 2.9% to a positive return of 5.2% in the four metal option.

The degree of influence exerted by individual factors represented by the variables in Table 3 reveals certain basic characteristics of an integrated nodule venture under existing economic conditions. First, project economics appear to be most sensitive to variations in metal prices. In Nyhart's model for example, sensitivity analysis of plus or minus 25% conducted on gross revenues reveals a return of over 13% for the ease of increased revenues and a return of 2.7% for the case of decreased revenues. By disaggregating the gross revenue into individual metal revenue streams, an additional insight obtained is the fact that real growth in metal prices is extremely influential in determining the economic returns to the project.

Table 2. Selected cost models for an integrated nodule venture (capital and operating costs)*

Variables		Models															
		Flipse (1980)		Nyhart (1980)		Hillman (1981)						Ingham (1985)					
						Venture 1			Venture 2			Australian Co.			Enterprise		
CAPITAL COSTS (in millions of US dollars—base year as in the model)																	
			%		%		%	%		%	%		%	%		%	%
1. Mining		324.7	32	352.6	31	538.7	33	30	521.4	33	30	528.5	35	30	528.5	35	30
2. Transportation		174.5	17	199.4	17	350.0	21	20	278.2	18	17	117.3	8	7	117.3	8	7
3. Processing	3M	522.0	51	588.9	52	757.0	46	—	757.0	49	—	845.1	57	—	845.1	57	—
	4M	—	—	—	—	886.3	—	50	886.3	—	53	1124.9	—	63	1124.6	—	63
TOTALS	3M	1021.2	100	1140.9	100	1645.7	100	—	1556.6	100	—	1490.9	100	—	1490.9	100	—
	4M	—	—	—	—	1775.0	—	100	1685.9	—	100	1770.7	—	100	1770.7	—	100

B

Variables		Flipse (1980)		Nyhart (1980)		Hillman (1981) Venture 1			Hillman (1981) Venture 2			Ingham (1985) Australian Co.			Ingham (1985) Enterprise		
ANNUAL OPERATING COSTS (in millions of US dollars—base year as in the model)																	
			%		%		%	%		%	%		%	%		%	%
1. Mining		77.3	34	68.1	31	71.5	33	22	70.1	33	23	89.4	36	19	93.1	36	19
2. Transportation		20.9	9	22.1	10	42.9	19	13	34.3	16	11	25.4	10	5	26.5	10	5
3. Processing: Energy	3M			47.5	22	35.9	16	—	35.9	17	—	40.5	16	—	40.5	15	—
	4M	—	—	—	—	84.8	—	27	84.8	—	28	159.1	—	34	159.1	—	33
Labour	3M	130.4	57	17.7	8	16.5	7	—	16.5	8	—	19.9	8	—	19.9	8	—
	4M	—	—	—	—	30.9	—	10	30.9	—	10	54.1	—	12	54.1	—	11
Misc.	3M			16.6	28	55.6	25	—	55.6	26	—	75.3	30	—	81.6	31	—
	4M	—	—	—	—	88.7	—	28	88.7	—	28	140.7	—	30	157.4	—	32
TOTALS	3M	228.6	100	217.0	100	222.4	100	—	212.4	100	—	250.5	100	—	261.6	100	—
	4M	—	—	—	—	318.8	—	100	308.8	—	100	468.7	—	100	490.2	—	100

* All models are for mining operations for 3 million dry metric tons/year except Nyhart which is for 2.72 million dry metric tons/year.

Table 3. Selected cost models for an integrated nodule venture (revenues)*

	Model					
	Flipse (1980)	*Nyhart (1980)*	*Hillman (1981)*		*Ingham (1985)*	
			Venture 1	*Venture 2*	*Australian Co.*	*Enterprise*
			Metal Prices (US$/lb)			
Nickel	3.75	3.75	3.50		3.24	
Copper	1.25	1.25	0.89		0.64	
Cobalt	5.50	5.63	7.00		11.70	
Ferro-Manganese	—	—	0.23		0.19	
			Total Company Revenues (US$)			
3 Metals	8450	9667	7563	8016	9 467	9 467
4 Metals	—	—	9615	9994	12 672	12 672
			Discounted Cash Flow Rate of Return (%)			
3 Metals	7.1[a]	9.21[b]	4.1[c]	6.0[c]	6.5[d]	7.8
4 Metals	—	—	3.5	5.2	(2.9)	0.07

* All models are for mining operations for 3 million dry metric tons/year, except Nyhart which is for 2.72 million dry metric tons/year.
[a] 100% equity financing, no depletion allowances, no payments are made to ISA, after tax estimates.
[b] 50% equity financing, tax and depletion allowances included, no payments to ISA.
[c] 100% equity financing, tax and depletion allowances included, no payments to ISA.
[d] 100% equity financing, no depletion allowances, payments to the ISA, before tax estimates.

This conclusion, when coupled with the trend observed during the past decade of generally depressed metal prices and investor lack of enthusiasm in participating in highly capital intensive and high-risk technological ventures, provides a partial explanation of the relatively low commercial interest in nodule mining projects at the present time.[7]

The second most sensitive variable in the project economics of an integrated nodule venture is capital costs. Technological problems may result in three different types of effects: efficiency effects, delays and cost effects. With regard to efficiency effects, two types are distinguished: input efficiency that may cause altered amounts of ore to be processed without commensurately increasing operating costs, and output tonnage efficiency effects which are not expected to be accompanied by cost effects.

As far as delays in the mining project due to technological problems there are two areas of uncertainty: the nodule mining system and the processing system. Although the processing system is considered the less risky of the two, its functioning depends on the existence of raw material, i.e. nodules are required to be available on a continuous basis and at specified quantities.

It has been argued by Johnson and Otto that if the lower risk processing stage could be separated from the higher risk mining sector, overall risks could be lowered.[8] They suggest the following options:

(1) design the processing plant so that at a modest incremental investment, it can be converted to process land-based nickel laterite ores;
(2) design the processing plant so that it can operate on blended nodule and laterite ores or on either nodule or laterite ores; and
(3) convert an existing facility for land-based laterite nickel ore to accept nodules.

NOTES

[1] United Nations, 1984. *Analysis of Exploration and Mining Technology for Manganese Nodules*. Seabed Mineral Series, Volume 2. Graham and Trotman (London), 140 pp.

[2] United Nations, 1986. *Analysis of Processing Technology for Manganese Nodules*. Seabed Mineral Series, Volume 3. Graham and Trotman (London), 79 pp.

[3] Flipse, J. E. 1983. "Deep Ocean Mining Economics" in Offshore Technology Conference, 2–5 May 1983 OTC 4491.

[4] Nyhart, J. D., Triantafyllon, M. S., Avaback, J., Blick, A., Sklar, B., Gillia, M., Kirkpatrick, D., Muggerridge, J., Nakagawa, R., Newman, J. and Will, A. 1983. *A Pioneer Deep Ocean Mining Venture*, MIT Sea Grant College Program, MITSG 83–14, 255 pp.

[5] Hillman, C. T. 1983. *Manganese nodule resources of three areas in the Northeast Pacific Ocean with proposed mining-beneficiation systems and costs*. Washington, D.C., US Department of the Interior, Bureau of Mines Information Circular 8933.

[6] Preparatory Commission for The International Sea-Bed Authority and for the International Tribunal for the Law of the Sea, 1986. The Enterprise: Economic viability of deep sea-bed mining of polymetallic nodules. (Document LOS/PCN/SCN.2/WP.10, 14 January 1986, prepared by Ingham, P. D.)

[7] C. J. Johnson and Otto, J. M. 1986. "Manganese nodule project economics: Factors relating to the Pacific region", *Resources Policy*, Vol. 12, No. 1, March 1986.

[8] C. J. Johnson and Otto, J. M. op. cit.

Chapter 4

LOCATION ANALYSIS FOR A PROCESSING SITE

The objective of site selection studies is to find a site that can be developed and from which operations can be conducted at the lowest overall costs with the greatest potential benefit in the long run. The investment in the development of a site can be amortized over a period of years along with the plant and buildings. A site that requires a greater initial expenditure for development may prove to be more economical in the long run because of consistent annual savings in other cost items than another site for which the initial expenditure is less.

The three elements of cost for location analysis are: (1) cost of obtaining suitable raw materials and supplies; (2) cost of conversion or production of the product for ready sale; and (3) cost of distributing the product to the dealer or consumer. Each of these cost elements will be affected by the location of the plant relative to the sources of raw materials and supplies, labour and equipment markets and markets for the finished product.

In the consideration of sites for the location of a manganese nodule processing plant in a developing country, both the investors and the authorities in the concerned countries have to take account of provisions within the United Nations Convention on Law of the Sea, that if judiciously used could be mutually beneficial and that appear to provide a means for shifting some of the processing capacity, discussed earlier, from industrialized countries to developing countries.

On the one hand, while the investor would be considering key factors such as market characteristics (especially if manganese is to be recovered from the processing operations), labour factors, infrastructure, financial incentives, fiscal considerations and political/legal risks, on the other, the authorities in the concerned developing countries would be considering key factors such as an improvement in the country's balance of payments, an increase in its foreign exchange earnings, transfer of technology, employment and training of its nationals and the possible forward and backward linkages between the new industrial concern and the country's economy.

The Convention on the Law of the Sea establishes a comprehensive framework for the regulation of all ocean space. It contains provisions governing, *inter alia*, the limits of national jurisdiction over ocean space, access to the seas, navigation, protection and preservation of the marine environment, utilization of living resources, seabed mining and exploitation of other non-living resources, and the settlement of disputes. It also has established a regime for the exploration and exploitation of minerals in the deep seabed area.

The Convention divides maritime space into two parts: areas falling under national jurisdiction (territorial sea, contiguous zone, exclusive economic zone, and the continental shelf) and area beyond national jurisdiction—the high seas and the international seabed area.

The Convention declares that "the Area and its resources are the common heritage of mankind". The body empowered to administer the common heritage of mankind and to regulate its exploration and exploitation will be the International Sea-Bed Authority (ISA), an international organization open to membership by all States as well as international organizations and other entities meeting specified criteria.[1] The Authority will have an Assembly, which will be the supreme body and will reflect the balance between the sovereign equality of all States, and a Council with limited representation. The Council will have primary responsibility over seabed mining activities, and will be advised by specialized commissions.

It is not the structure but the functions of the International Sea-Bed Authority which make it a forerunner in the development of the law of international organizations. All mining activities in the Area are under the control of the International Sea-Bed Authority.

A parallel system is established under which the Authority is empowered both to undertake mining operations through its commercial arm—the Enterprise, and to enter into contracts enabling private and State ventures to acquire mining rights.

The conditions and modalities for financing the Enterprise and for ensuring that it is technologically equipped to carry out activities form an integral part of the package.[2] The Preparatory Commission for the International Sea-Bed Authority and for the International Tribunal for the Law of the Sea created by resolution I of the Third United Nations Conference on the Law of the Sea, has also been entrusted to carry out functions under resolution II, which governs preparatory investment in pioneer activities. Under this resolution, certain protections are granted to qualifying seabed miners who apply to the Commission and are registered by it to conduct exploratory activities. It is the Commission which is empowered to fulfil certain functions on behalf of the international community as a whole.

Again, although there is no specific obligation in the Convention for the investor to provide training and/or to transfer technology in respect of the processing stage of a manganese nodule venture, considerations that go beyond the traditional concerns in siting a processing facility can be brought to bear on the decision. Aside from those noted above, there are additional elements that modify the nature of the decision-making process. On the investor's side is the fact that the source of nodules is not a land-based mine in the country in question but the international Area, and on the developing country's side is the fact that nodule processing will be energy intensive and will generate the need for a significant amount of waste disposal.

INVESTOR CONSIDERATIONS REGARDING THE PLANT LOCATIONS DECISION

Where to locate a new plant is one of the key strategic questions for any company. The plant represents a major investment in physical and human assets which changes the geographic spread of the company. The investment often involves considerable risk. Although a new plant has the potential to operate efficiently, it may be many years before cost targets can be realized. Under actual operating conditions, assumptions about production logistics may prove to be invalid. If plant performance is unsatisfactory, withdrawal or liquidation is costly and difficult. The consequences of choosing a bad location that puts a new plant in the wrong place with resultant high operating costs, would remain for many years.

In determining the desirability of selecting one location over another, the investor will examine the key factors that influence the rate of return on investment. These factors include: market characteristics, financial incentives, fiscal considerations, labour factors, infrastructure characteristics, environmental regulations, and political factors. Investors will also consider short- or long-term influences on these factors.

Market Characteristics

Market characteristics affect the demand and price for the products of the new plant. This includes such considerations as the total size of the market and its rate of growth. This also includes the pricing in the market and the movement of price cycles. The output from a nodule processing plant would likely be priced at world prices based on some metal exchange such as the London Metal Exchange or at a producer price. Nickel is in general priced on a producer price basis, copper is priced on the metal exchange, and cobalt and manganese are priced by individual negotiations. This pricing mechanism is no different from land-based mining operations. Investors are likely to act to reduce risk by selling the product on a long-term contractual basis. However, plant location would not be greatly altered as a result of market considerations, except in the case of manganese. Copper, nickel and cobalt are high-valued concentrated products. They can be shipped in small ships. Manganese, on the other hand, is a bulky product and ship size and loading possibilities would affect the location of markets and the price received by its nodule producer.

In the investor's analysis of market characteristics a factor that could significantly alter its decision, would be if there was either a producing or a proven laterite nickel orebody within the vicinity of the processing plant.[3] This would minimize the very significant risk that the investor in the processing plant would face from lack of performance in the deep sea mining component of the project. In this instance, it will be expected that the investor would select the option of designing the processing plant so that at a modest incremental investment, the plant could be converted to process lateritic nickel ore.

Financial Incentives

Investors are concerned about financial incentives because these affect their rate of return through the increase in revenue to the mining company. Two types of incentives are possible. First there are the incentives that may be offered by the International Sea-Bed Authority (commonly referred to as the Authority), and second there are the incentives that may be offered by the authorities in the developing State or States.[4] The former are still under negotiation within the context of the Preparatory Commission for the International Sea-Bed Authority and for the International Tribunal for the Law of the Sea and as a result cannot be incorporated in this analysis. The latter, however, have a long history and will be presented in this study.

The direct incentives that may be offered by the government of the host country in an effort to encourage investment include, *inter alia*, investment grants, low cost loans and guarantees of repayment to lenders outside the host country.

Often, such incentives are coupled with fiscal concessions such as permitting large remittances with low tax.

Investment incentives come in many forms and their payment may be made at varying times during the project's life. Initial capital investment is often stimulated by the use of outright cash grants towards the cost of the plant, land or buildings or infrastructure that may be required by the metallurgical plant. This infrastructure includes roads, power lines, communication systems, town sites and other similar investments. On occasion, such incentives can take the form of free or low cost buildings and land provided to the company's specifications. Alternatively, low or zero-interest rate loans may be provided. Further incentives may be available once the project has commenced operation, such as special grants to cover employee training, wage subsidies, and rapid depreciation allowances to increase rate of return through increased cash flow.

The timing of investment incentives is especially important. At the beginning of investment in a project, incentives in cash or as free plant and equipment, or other infrastructure will usually be substantially more valuable than depreciation allowances or tax holidays on profits unless the company has other profit-making operations against which the allowances may be used. The stability of incentives is a further point that is examined carefully. There is no point investing in a new location on the expectation of future incentives only to find that by the time these come to fruition, the original basis has been changed.

Fiscal Considerations

Another area for investigation by investors in comparing countries is that of taxation. This includes both corporate and personal, direct and indirect tax rates, tax treatments on dividends, interest payments, technology, management fees and the like, tax treatment for depreciation and other expenses, treatment of losses and rates of excise duty.

Tax concessions are an important form of investment incentive. Local tax holidays are a common form of incentive. These are especially useful for low capital intensity firms which expect to show high profits quickly from their investments.

In addition to tax holidays, many countries operate customs-free zones. These are of less use to mining firms since the product is often produced internally. However, customs-free zones are useful to reduce the cost of imported supplies used in the plant or in auxillary facilities.

Labour Factors

The availability and quality of labour are increasingly important determinants of location. In reviewing these considerations, the first factor which must be checked is that adequate labour is available in the number of people and the mix of skills required. Secondly, since the level of know-how available locally is often less than what is required to operate successfully, the mining company should be prepared to introduce a substantial amount of job training and in some cases basic literacy education. It may also be necessary in many cases to use local nationals in the majority of the managerial positions. Again, in many developing countries there is a serious shortage of skilled managers. The investor may therefore need to develop extensive educational programmes in order to develop the necessary technical and linguistic skills.

In addition to availability and quality of labour, relations between companies and unions would be carefully investigated. The structure and attitude of the local trade unions, their record for restrictive practices, unofficial disruptions and strikes, and their ability to maintain satisfaction amongst their members are important considerations in the choice of a location.

Labour law and political customs in many countries also tend to impose increasing restrictions on business decisions. Employment protection legislation in many countries also tends to restrict the right of employers to hire or fire or establishes expensive rules for severance pay. Finally, the social security and fringe benefit requirements laid down by legislation in many countries may increase the cost of labour substantially.

Labour costs are also an important consideration. The rising cost of labour in the developed countries has led to the widespread growth of satellite plants in less developed countries with ample supplies of relatively cheap labour. In a study conducted under the sponsorship of the University of Hawaii Sea Grant College Program through a grant from the National Oceanographic and Atmospheric Administration (NOAA) Office of Sea Grant, entitled "Analysis of Laws Applicable to Manganese Nodule Processing and Investment Climates in Representative Pacific Areas", and on the subject of labour costs in six Pacific basin-rim countries (Australia, Canada, Fiji, Ecuador, Colombia and the Philippines), it was concluded that "... the Canadian and Australian labour costs are more than six times as high as Fiji, Ecuador, Colombia and the Philippines. The Philippines has the lowest labour costs at substantially less than $1.00 per hour. Fiji, Ecuador and Colombia have average labour costs in manufacturing in the $1.00–$2.00 per hour range and Australia and Canada have a mining industry average salary of $14.00–$16.00 per hour. With respect to longer-term wages that might apply in a nodule processing plant labour wages in the $2.00–$4.00 per hour range in constant 1984 dollars are believed to be realistic for Fiji, Ecuador and Colombia. The Philippines is expected to continue to have the lowest labour costs at less than $2.00 per hour in constant 1984 dollars".[5]

Infrastructure Characteristics

It is important to consider the infrastructure of each area in the selection of a site for a nodule processing plant. The key points that would be checked include, *inter alia,* transport facilities, port facilities, ease of exporting and importing, availability of local supplies of raw materials and packaging, availability of energy supplies, geographic distance between the proposed plant site and suppliers and consumers, and ease of obtaining permission for the various activities.

The infrastructure would also be considered from the viewpoint of key employees and their families, both at the country and individual site levels. The availability of suitable housing, educational facilities, social and recreational facilities, the type of climate, crime rate and so on will be important influences in attracting executives and their families and are also significant for labour relations and employee motivation.

With regard to transport facilities a prerequisite is an existing or prospective port site with a potential draft of 12.2 m and sufficient space to handle 60 000 dead weight ton (DWT) or larger ore carriers. If a pier with off-loading equipment is unavailable, then a water frontage area will be required for this purpose as well as for a surge storage area bunkering and ship service facilities.

Two basic scenarios have been considered in studies on the assumed availability of plant sites and their relative proximity to a port that meets the requirements noted above. Studies by NOAA have assumed that plant sites on the Pacific or Gulf coasts would be 40 to 100 km from the port.[6] Canada Metal's (CANMET) studies of British Columbia have nominated plant sites that are within 2 to 12 km of the unloading docks.[7] Studies for the State of Hawaii have considered plant sites from 3 to 16 km from the port.[8]

Nodules may be transported from the port to the plant site in one of three ways: rail, conveyor and slurry pipeline. Since the internal transportation of nodules from port to plant site adds nothing of value to the product, its cost, if excessive, would render a site economically non-competitive.

Port facilities for the receipt of inbound nodules and all bulk materials (coal, limestone and silica sand) will be required. If the processing plant is contiguous or within 3 km of the port, it is unlikely that any intermediate storage and recovery facilities would be required. If the plant were located at a greater distance, intermediate storage and recovery facilities as well as an overland haulage system would be required.

It is expected that nodules will be transferred from the mineship into transport vessels that would then carry nodules to the marine terminal. The numbers and sizes of such vessels are a function of the number of mineships, the mining rate, minesite to port distances, draft limits at the port, loading and unloading rates, and transport speed. When larger vessels are used, fewer are needed than when smaller ones are used and this could reduce cost. Thus, industry is expected to use the smallest number of the largest sized vessels. For the US West Coast, these ships could be as large as 60 000–70 000 DWT.[9]

In optimizing the fleet, however, consideration will be given to redundancy in case of breakdowns, to provide buffer capacity to allow for dry dock maintenance and operational delays and to provide additional transport space during periods when production rates by the mineship are unusually high.

It has also been proposed that barges could be used as transport vessels.[10] Under one concept, a tug would bring out an empty barge, drop it off and return to shore with a loaded barge. The tug-barge combination could be either a hawser-towed barge system, or an integrated tug-barge system with a firm mechanical connection between the tug and barge. If technically feasible, the use of tugs and barges could result in savings in marine transportation costs.

Tables 4 and 5 summarize the characteristics of various ports (size, depth at wharf or pier) and their distances from the areas of maximum commercial interest. In the Pacific Ocean, this is the Deep Ocean Mining Environmental Study Group's B mine site (Fig. 1) and in the Central Indian Ocean basin, it is the recently registered pioneer area (Fig. 2). For illustrative purposes, Tables 6 and 7 estimate nodule transport requirements from these two sites to those ports with a draft of 12.2 m and with sufficient space to handle 60 000 DWT or larger ore carriers.

In estimating nodule transport requirements, a number of salient factors relating to the moisture content of nodules, transport vessel size and hull characteristics and facilities have to be taken into account. Typical ore carriers that have been discussed in the literature on nodule transport necessitate a minimum depth at wharf or pier at the port to be 12.2 m. It is entirely possible to modify the hull form of an ore carrier to yield a displacement of less than 12.2 m. For this reason, a number of ports are included in Tables 4 and 5 whose maximum allowable displacement is less than 12.2 m. If such ports are to be used either they must be deepened or preferably the hull form of the ore carriers has to result in a displacement that is within the maximum permitted at these ports.

Nodules *in situ* will have a moisture content of 70%. When slurry water is removed enroute to port through a decant and bottom drainage piping on board the vessel, the moisture content is reduced to 40%. Loading and unloading times found in the literature vary with the various assumptions used. In general, for the type of transport vessel referred to above, loading and unloading times have each been estimated in the range of 28–32 hours. For the purpose of this illustration, they were each assumed to be 30 hours. Finally, the average speed of the transport vessels was taken to be 16 knots.

Transshipment from Port to Processing Plant[11]

Facilities for the receipt of inbound nodules, nodule storage and start of inland transportation would be needed at any processing plant. Terminals may be designed for handling of nodules in 3 main forms: slurry, whole, or dried/ground. While terminals designed to unload nodules mechanically or pneumatically are possible, terminals designed to unload nodules in a slurry form are currently considered most likely.

Slurry Terminals

The primary features of a slurry terminal are a pier for tying up the nodule (crushed or whole) transport ships, one or more travelling cranes on the pier for swinging portable pumps onto each transport ship, containment areas to hold nodules awaiting onshore transportation to the processing plant, tailings holding facilities (only if ocean disposal is used), and a facility for

Table 4. Estimated distances from the most likely mining site in the Pacific to potential ports for processing sites.

Port	*Size*[a]	*Latitude*	*Longitude*	*Depth at wharf or pier*[b] *(m)*	*Distance*[c,d] *to point A (naut. miles)*
(1) *NORTH AMERICAN WEST COAST (including CANADA and HAWAII)*					
USA					
(1) Seattle	L	4735N	12220W	Over 23.2	2300
(2) Portland	L	4531N	12240W	12.5–13.7	2184
(3) Los Angeles	L	3400N	11815W	17.1–18.2	1734
(4) San Francisco	L	3745N	12227W	11.0–12.2	1782
Hawaii					
(1) Hilo	M	1942N	15504W	9.4–10.7	1075
(2) Honolulu	M	2119N	15750W	11.0–12.2	1256
Canada					
(1) Vancouver	L	4913N	12306W	14.0–15.2	2376
(2) Prince Rupert		5418N	13017W		2585
(2) *WEST COAST OF CENTRAL AMERICA (MEXICO–PANAMA)*					
Mexico					
(1) Acapulco de Juarez	S	1651N	09956W	11.0–12.2	2254
(2) Manzanillo	S	1900N	10420W	11.0–12.2	2015
Guatemala					
Puerto de San José	V	1400N	09050W	7.9–9.1	2782

(3) *SOUTH AMERICAN WEST COAST*					
Colombia					
Cartagena	S	1024N	07533W	11.0–12.2	3694
Peru					
Callao	M	1205S	07708W	11.0–12.2	3918
Chile					
Arica	S	1830S	07020W	11.0–12.2	4415
(4) *AUSTRALASIA*					
Australia					
(1) Melbourne	L	3745S	14458E	11.0–12.2	5211
(2) Newcastle	L	3255S	15146E	11.0–12.2	4802
(3) Sydney	L	3355S	15110E	11.0–12.2	4851
Tasmania					
Hobart	M	4254S	14718E	14.0–15.2	5207
New Zealand					
(1) Auckland (North Island)	L	3655S	17447E	11.0–12.2	3933
(2) Port Lyttleton (South Island)	M	4336S	17242E	11.0–12.2	4257
(3) Wellington (South Island)	L	4117S	17447E	11.0–12.2	4098
(5) *SOUTHEAST ASIA ARCHIPELAGO*					
Fiji					
Suva	S	1808S	17825E	12.5–13.7	3121
Papua New Guinea					
Port Moresby	S	0930S	14707E	7.9–9.1	4620

Table 4 continued

Port	*Size*[a]	*Latitude*	*Longitude*	*Depth at wharf or pier*[b] *(m)*	*Distance*[c,d] *to point A (naut. miles)*
Indonesia					
(1) Surabaya	M	0714S	11245E	11.0–12.2	6590
(2) Sambu	S	0106N	10354E	12.5–13.7	7009
Philippines					
Manila	L	1437N	12058E	11.0–12.2	5825
(6) *NORTH EAST ASIA*					
Republic of Korea					
Pusan	L	3505N	12902N	11.0–12.2	5122
Japan					
(1) Kawasaki	L	3532N	13941E	20.1–21.3	4602
(2) Yokohama	L	3528N	13938E	15.5–16.8	4605

[a] *Size*. The classification is based on several applicable factors, including area, facilities and wharf space. It is not based on area alone or on any other single factor. L = large; M = medium; S = small; V = very small.
[b] *Depth (wharf/pier)*. The greatest depth at chart datum alongside the respective wharf/pier is given in metres.
[c] Point A has the following coordinates 11°42N, 138°24W.
[d] Distances calculated by Robert McKay (UN Cartographic Unit).
Source: World Port Index. Pub. 150, Ninth Edition 1984. Published by the Defense Mapping Agency. Hydrographic/Topographic Center.

Table 5. Estimated distances from the most likely mining sites in the Central Indian Basin to potential ports for processing sites

Port	*Size*[a]	*Latitude*	*Longitude*	*Depth at or pier*[b] *(m)*	*Distance*[c, d] *to point B (naut. miles)*
(1) *ASIA*					
India					
(1) Vishakha-patnam	M	1742N	08324E	9.4–10.7	1893
(2) Madras	M	1305N	08018E	9.4–10.7	1586
(3) Bombay	L	1856N	07251E	9.4–10.7	1925
Malaysia					
(1) Pelabuhan Kelang	M	0257N	10124E	9.4–10.7	1790
(2) Penang	M	0530N	10028E	9.4–10.7	1833
Sri Lanka					
(1) Colombo	L	0657N	07952E	9.4–10.7	1217
(2) Trincomalee	M	0834N	08113E	12.5–13.7	1331
Indonesia					
(1) Airbangis	V	0009N	09925E	11.0–12.0	1601
(2) Sambu	S	0106N	10354E	12.5–13.7	1894
(3) Surabaya	M	0714S	11245E	11.0–12.2	2196
(2) *AFRICA*					
Malagasy Rep.					
Tamatave (Toamasina)		1810S	4923E		1568
(3) *REPUBLIC OF SOUTH AFRICA*					
(1) Durban		2953S	3100E		2692
(2) Port Elizabeth		3358S	2536E		3009

[a] *Size*. The classification is based on several applicable factors, including area, facilities and wharf space. It is not based on area alone or on any other single factor. L = large; M = medium; S = small; V = very small.

[b] *Depth (wharf/pier)*. The greatest depth at chart datum alongside the respective wharf/pier is given in metres.

[c] Point B has the following coordinates 13°S, 76°E. The distance computed is the great circle distance to the nearest nautical mile.

[d] Distances calculated by Robert McKay (UN Cartographic Unit).

Source: World Port Index, Pub. 150, Ninth Edition 1984. Published by the Defense Mapping Agency. Hydrographic/Topographic Center.

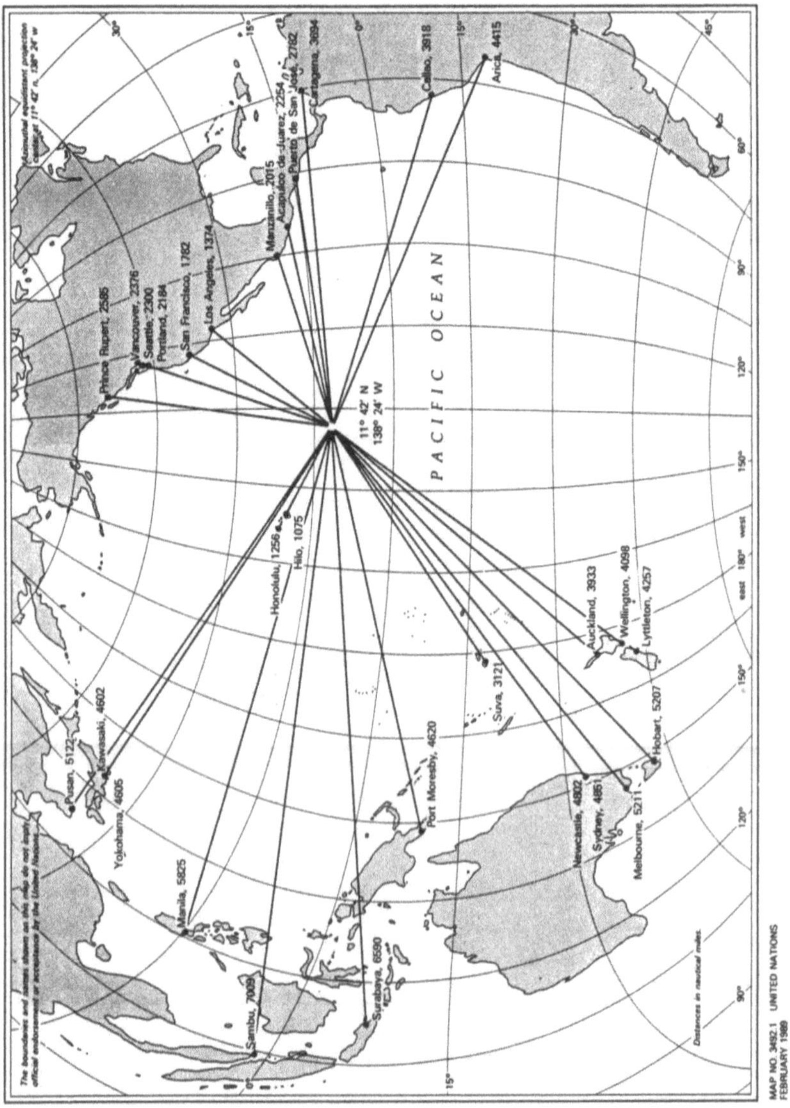

Fig. 1. Distances from the most likely site in the Pacific to potential ports for processing sites.

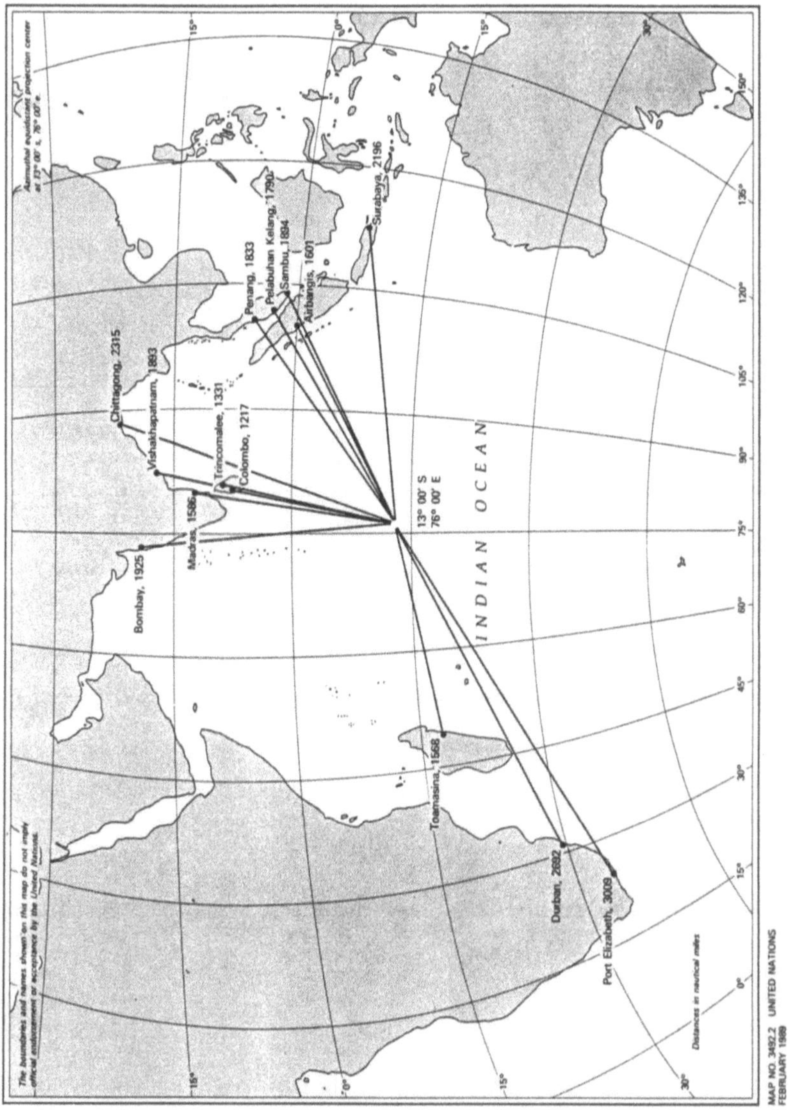

Fig. 2. Distances from the most likely site in the Central Indian basin to potential ports for processing sites.

Table 6. Nodule transport requirements from various ports bordering the most likely mining sites in the Pacific

Port	Distance to site (naut. miles)	Days per round trip	Round trips[a] per year	Million tonnes[b] per year	Calculated vessels required[c]
Los Angeles	1734	11.5	29	1.5	3
Portland	2184	13.9	24	1.2	4
San Francisco	1782	11.8	28	1.4	3
Seattle	2300	14.5	23	1.2	4
Hilo, Hawaii	1075	8.1	41	2.0	2
Honolulu, Hawaii	1256	9.0	37	1.8	2
Vancouver	2300	14.5	23	1.2	4
Prince Rupert	2585	16.0	21	1.1	4
Acapulco	2254	14.2	23	1.2	4
Manzanillo	2015	13.0	25	1.3	5
San José	2782	17.0	19	1.0	5
Suva	3121	18.8	18	0.9	5
Kawasaki	4602	26.5	12	0.6	7

[a] Round trips are rounded off to the nearest integer.
[b] Based on the following:
(1) Annual tonnage required (3×10^6 dry tons per year) with a moisture content of 30% (4.35×10^6 wet tons per year).
(2) Vessels of capacity 55 000 dwt carry 50 000 wet tons per trip.
(3) Average speed of vessel = 16 knots.
(4) Load and discharge times each 30 hours.
(5) 330 days/year of service.
[c] Number of vessels are rounded off to the nearest integer.

loading fuel destined for the mineship by way of returning transport vessels (Fig. 3).

Dry Nodule Terminals

Nodules could also be shipped to shore in either a dry whole bulk form or in a dried and ground form. In either case, the terminal facilities would be slightly different than described above for a slurry system. An illustration of a terminal for dry whole nodules is presented in Fig. 4. A marine terminal for

Table 7. Nodule transport requirements from various ports bordering the most likely mining sites in the Central Indian Ocean Basin

Port	*Distance to site (naut. miles)*	*Days per round trip*	*Round trips[a] per year*	*Million tonnes[b] per year*	*Calculated vessels required[c]*
Vishakhapatnam	1893	12.3	27	1.3	4
Madras	1586	10.7	31	1.5	3
Bombay	1925	12.5	26	1.3	4
Pelabuhan Kelang	1790	11.8	28	1.4	4
Penang	1833	12.0	28	1.4	4
Colombo	1217	8.8	37	1.9	3
Trincomalee	1331	9.4	35	1.8	3
Surabaya	2196	13.9	24	1.2	4
Sambu	1894	12.4	27	1.4	4
Airbangis	1601	10.8	31	1.6	3
Durban	2692	16.5	20	1.0	5
Port Elizabeth	3009	18.2	19	1.0	5
Tamatave (Toamasina)	1568	10.7	31	1.6	3

[a] Round trips are rounded off to the nearest integer.
[b] Based on the following:
(1) Annual tonnage required (3×10^6 dry tons per year) with a moisture content of 30% (4.35×10^6 wet tons per year).
(2) Vessels of capacity 55 000 dwt carry 50 000 wet tons per trip.
(3) Average speed of vessel = 16 knots.
(4) Load and discharge times each 30 hours.
(5) 330 days/year of service.
[c] Number of vessels are rounded off to the nearest integer.

dried and ground nodules would probably resemble a terminal designed to handle dry bulk chemicals, fertilizers, or cement (Fig. 5).

Enclosed handling and storage equipment would be required for dust control and to reduce loss of material. Since 30–40% of water weight of nodules would be lost during shipboard nodule drying, either lower cargo tonnages could be handled or fewer vessels would be needed and terminal arrivals would be less frequent. The essential elements of a dried and ground nodule terminal are the ship dock, the pier or wharf,

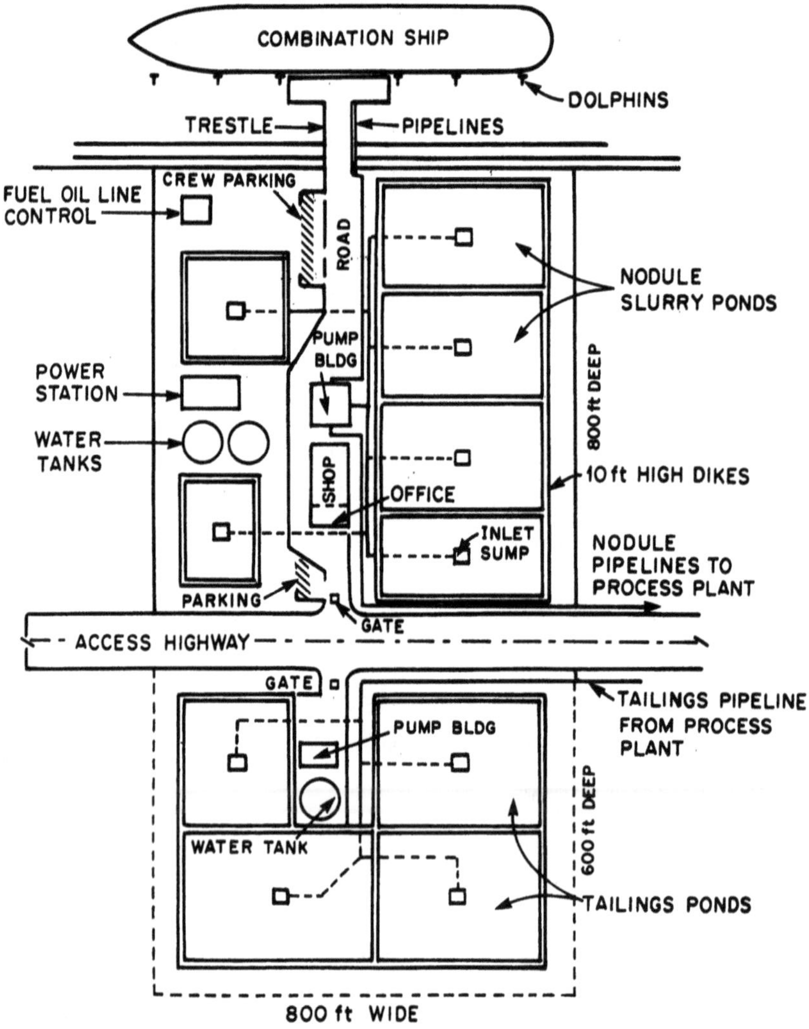

Fig. 3. Minimal slurry terminal with alternative extension for tailings storage (source: US Department of Commerce, "Gulf Coast Manganese Nodule Processing Plant Location Criteria").

shoreside or shipboard centrifugal suction cargo unloaders, or clamshell or bucket unloaders and an enclosed conveying system to a stockpile building (Fig. 4).

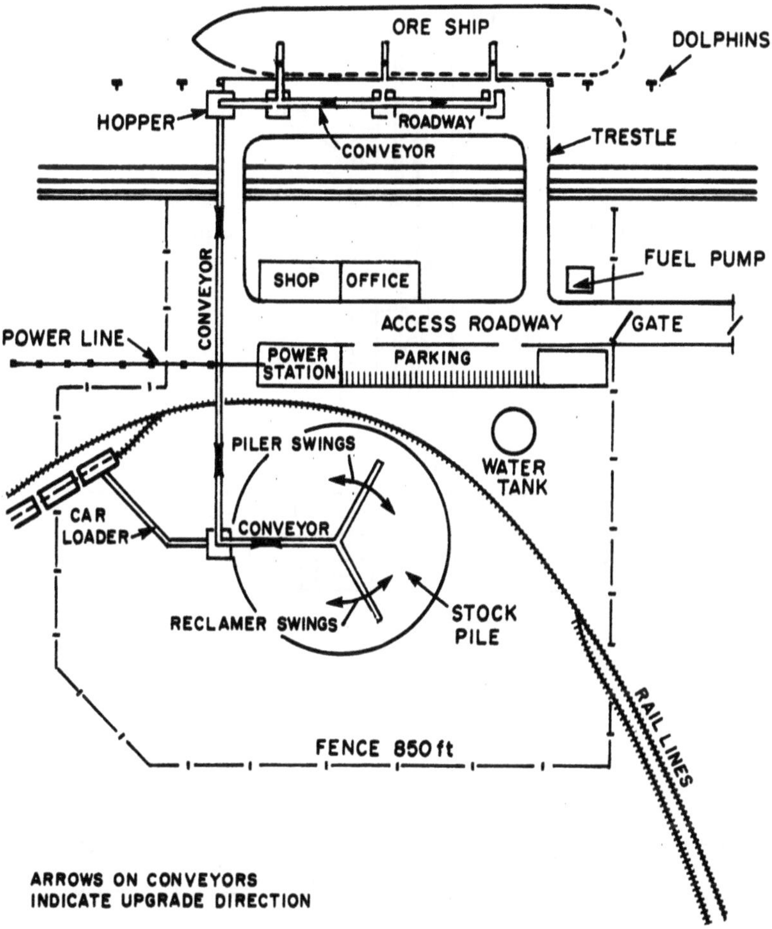

Fig. 4. Minimal whole nodule terminal (source: US Department of Commerce, "Gulf Coast Manganese Nodule Processing Plant Location Criteria").

Port to Plant Transport

Land transportation of nodules from port terminals to processing plants and the transport of waste to disposal areas would depend, to a large extent, on the physical form of the

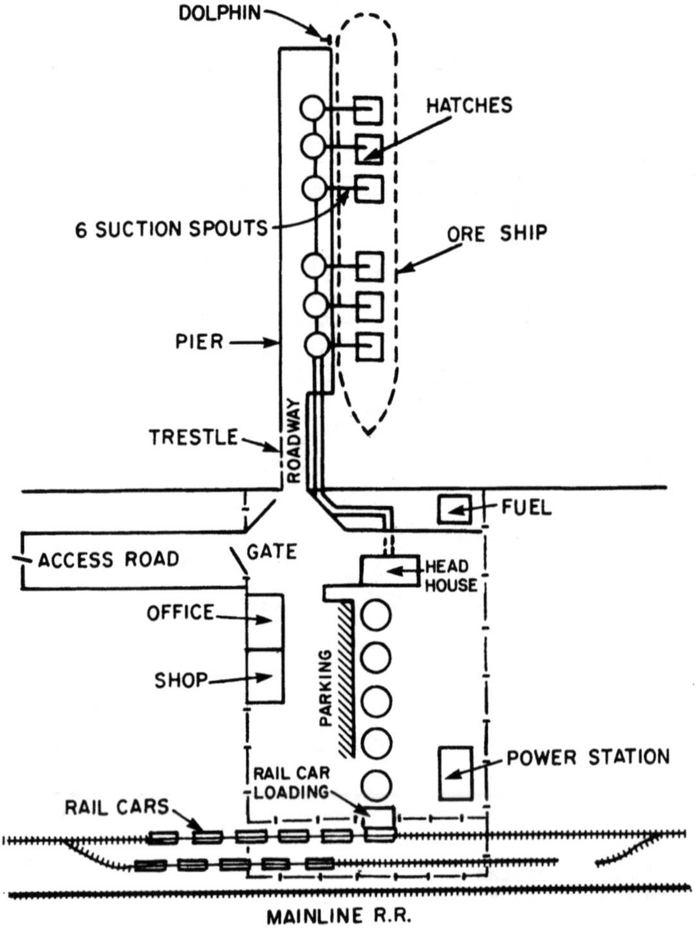

Fig. 5. Minimal dried nodule terminal (source: US Department of Commerce, "Gulf Coast Manganese Nodule Processing Plant Location Criteria").

nodule material. If the material is brought to port wet, either as broken nodules or crushed and ground, it could be conveyed to the processing plant in *a slurry pipeline*. If the nodules are ground and dried, either *conveyor* or *rail transportation* appears most likely.

Resource Requirements of Three and Four Metal Processing Plants

Resource requirements vary widely between three and four metal plants. The latter may require three to four times as much electricity as the former but only one-fifth as much land for onshore waste disposal. This is as a result of the different metal recovery processes used and also the fact that while nickel, copper and cobalt together make up only 3% of nodules by dry weight, manganese amounts to 25%. Given the size of the near-term manganese market, a four metal plant will probably be designed to process about one-third as many nodules as a three metal plant.

The principal differences in resource requirements for onshore processing for three and four metal plants are presented in Table 8. While a three metal plant would be designed to recover copper, nickel and cobalt as its primary products, a four metal plant would also produce manganese as a primary product.

If recovery of only three metals is desired, the reduction of manganese must be carefully controlled. Not only is chemical reduction an energy intensive and expensive step, but it would be highly desirable not to further complicate the required nickel/copper/cobalt separation steps with the presence of dissolved manganese if it could be maintained as a relatively inert, benign oxide.

A major factor in determining whether or not to produce manganese will be the near-term market for this metal. The percentage of manganese (25% on average) in nodules is such that a single nodule processing operation recovering 1 million (dry) tons of nodules per year could be expected to supply about one-third of the annual requirements of the United States. It is currently assumed that a first generation four metal plant would be designed to process only 1 million tons of nodules, dry weight.

In respect of resource requirement differences between three and four metal plants, Table 8 reveals the following differences.

Table 8. Preliminary identification of resource requirements for major activity alternatives for onshore marine nodule processing[a]

Major activities	*Resource requirements*															
	Three metal processing								*Four metal processing*							
					Energy								*Energy*			
						Fuels								*Fuels*		
	Nodule Input (10^6 t/yr)	Land (hectares)	Labour (persons)	Water (10^6 m^3/yr)	Electrical (10^3 kw)	Coal (10^3 t/d)	Gasoline (10^4 l/yr)	Diesel (10^4 l/yr)	Nodule Input (10^6 t/yr)	Land (hectares)	Labour (persons)	Water (10^6 m^3/yr)	Electrical (10^3 kw)	Coal (10^3 t/d)	Gasoline (10^4 l/yr)	Diesel (10^4 l/yr)
(1) *Transshipment From Port to Processing Plant*																
• Marine Terminal																
Whole Nodules	—	8.0	58	—[b]	N/A	—	—	—	—	4.8	49	—	N/A	—	—	—
Slurry	—	5.2	33	N/A	N/A	—	—	—	—	2.4	28	N/A	N/A	—	—	—
Dried/Ground	—	4.4	47	—	N/A	—	—	—	—	3.2	39	—	N/A	—	—	—
• Transportation System																
Rail[c]	—	N/A	15	—	N/A	—	—	56.0	—	N/A	12	—	N/A	—	—	19.0
Truck	—	N/A	N/A	—	—	—	N/A	N/A	—	N/A	N/A	—	—	—	N/A	N/A
Slurry[d]	—	4.8+ 3.8/km	6–12	2.70	1.2	—	N/A	N/A	—	2.4+ 3.8/km	6–12	0.81	0.4	—	N/A	N/A
Conveyor[e]	—	4.8+ 3.2/km	N/A	—	0.5	—	N/A	N/A	—	1.6+ 3.2/km	6	—	0.5	—	N/A	N/A

(2) *Nodule Processing at Onshore Plant*																
• Reduction/Ammonia Leaching Plant	3.7	80.0	500	6.00	23.5	2.1	—	—	—	—	—	—	—	—	—	—
• Cuprion/Ammonia Leaching Plant	3.7	80.0	500	7.90	0.0	1.9	—	—	—	—	—	—	—	—	—	—
• High Temperature Sulfuric Acid Plant	3.7	73.0	500	5.40	28.8	1.0	—	—	—	—	—	—	—	—	—	—
• Reduction/HCl Leaching Plant	—	—	—	—	—	—	—	—	1.4	40.0	300	1.40	94.5	1.2	—	—
• Smelting Plant	—	—	—	—	—	—	—	—	1.4	50.0	300	1.40	70.3	0.8	—	—
(3) *Waste Disposal*																
• On-site and/or Off-site Disposal																
Landfill	—	20.0/yr	20	0.09	0.23	1.5	8.7	160.0	—	4.0/yr	5	0.02	0.12	N/A	2.7	27.0
Tailings Pond	—	40.0/yr	4	Neg.	0.66	—	N/A	N/A	—	7.2/yr	3	Neg.	0.06	—	N/A	N/A
Manganese Storage	—	N/A	N/A	N/A	N/A	N/A	N/A	N/A	—	N/A	N/A	N/A	N/A	N/A	N/A	N/A
• Off-site Disposal																
Ocean Dumping	—	—	N/A	—	N/A	—	—	N/A	—	—	N/A	—	N/A	—	N/A	N/A

Abbreviations: m^3/yr: cubic metres per year; t/d: tons per day; KW: Kilowatt; N/A: data not available; 1/yr: litres per year; km: kilometre; Neg.: negligible.

[a] All values are rounded; all units are metric

[b] —indicates none

[c] Assumes: 8 km of rail

[d] Assumes: 32 km pipeline

[e] Assumes: 3050 metre length, 24 hour continuous operation

Source: Deep Sea Mining: Final Programmatic Environmental Impact Statement. Vol. 7, Department of Commerce, NOAA, September 1981.

Energy

A major input will consist of a combination of hydrocarbon fuel and purchased electrical power. It is generally expected that coal will be the hydrocarbon fuel chosen given the current uncertainties with oil and gas.[12] Depending on the processing route selected, coal usage could range from 900 to 2140 metric tons per day. Full onsite generation is possible but from the operator's point of view may only be a last resort since it would be an inefficient use of his capital. Quantities sought from outside the plant would range from about 25 megawatts per year for a first generation three metal plant to nearly 100 megawatts for a first generation four metal plant.

Fresh Water

Fresh water will be used for the generation of steam, cooling, processing and perhaps in the slurrying of wastes. For steam generation and cooling, there are limits on water hardness and suspended solids in order to avoid process system foul-up. Although this would have to be fresh water, it would not need to meet drinking standards. The quantities of water required would range from 4 million to 24 million litres per day depending on three or four metal plant designs. While cooling water is expected to be recycled, some process water will be lost as waste.

Land

Since copper, nickel and cobalt account for only 3% of nodules (dry) weight, solid waste in a three metal plant will amount to approximately 97% of nodules weight. For a four metal plant, the recovered metals amount to about 30% and solid waste 70% of the dry weight. Thus whether three or four metals are recovered, a sizeable percentage of the nodules will be unused. A relatively large and flat land will be required for tailings ponds. A three and four metal plant would require approximately 40 hectares and 8 hectares per year respectively to 13 m depth for tailings disposal over the 20 years (or more) operating period of a processing plant.

Although the types and quantities of reagents will probably vary from process to process, the combination of the major materials used directly in nodule processing is summarized in Table 9. It is anticipated that most reagents will be shipped to

Table 9. Composition of process materials and supplies

GASES
(1) Ammonia—Commercial anhydrous
(2) Hydrogen—Commercial, 99% minimum
(3) Hydrogen sulphide—(Liquid). Commercial, 97.5%
(4) Chlorine—(Liquid). Commercial, 99.5%
(5) Nitrogen—Commercial

LIQUIDS
(1) Organic—Liquid ion exchange/chelating agent dissolved in dilutant at concentrations appropriate to each process
(2) Sulphuric acid—Commercial, 93%
(3) Nitric acid—Commercial, 60%
(4) Sodium hydroxide—Commercial, 50%
(5) Fuel—vehicular and combustion fuel as required
(6) Oxygen—Commercial
(7) Water—Salt and fresh

SOLIDS
(1) *Calcium* in *Limestone* (Calcium carbonate, 80% calcium) and *Lime* (Calcium oxide, 79% calcium)
(2) *Magnesium* in *Limestone* (Magnesium carbonate, 15% magnesium) and *Lime* (Magnesium oxide, 12% magnesium)
(3) *Inerts*, 5% in *Limestone*, 9% in *Lime*
(4) Flocculants—Commercial polyelectrolytes
(5) Additives—Commercial electrowinning additives
(6) Sodium Sulphate—Anhydrous, photo grade
(7) Boric acid—Commercial granular, 99.9%
(8) Carbon—Commercial activated carbon
(9) *Water treatment chemicals*
Borax—Technical, anhydrous, 99%
Electrode paste
Salt—Commercial rock salt

ENERGY NEEDS
Coal—Generated electricity

Source: Deep Sea Mining: Final Programmatic Environmental Impact Statement. Vol. 7, Department of Commerce, NOAA, September 1981.

the plant in commerical concentrations by bulk transportation methods. The exceptions would be those reagents which are particularly hazardous or which could be easily generated on site, e.g. hydrogen sulphide.

A final item of importance to site selection is general and administrative support, that is: taxes (local, state), insurance, labour relations, wage rates, community consideration (cost of living, schools), climate, communications, and government (local, state). The state and local tax situation must be evaluated. In addition, the nature of the encouragement and incentives that the government gives to businesses looking to site plants in the area under its jurisdiction will also be evaluated.

Environmental and Land-Use Laws

Despite the amount of environmental research and laboratory scale tests undertaken so far, until large-scale, long-term activities take place at sea and onshore, significant environmental uncertainties will continue to exist. Environmental perturbations are expected to occur during the mining of nodules and again, when they are processed.

With respect to the environmental effects of mining, research work on two of the systems that have been developed, namely the hydraulic mining system and the continuous line bucket system, reveal that to varying degrees both methods will have potential environmental effects in the near-surface and near-bottom ocean environments. It is expected that the near-bottom zone will be affected by the shallow dredging operation (upper half metre) and the consequent resettlement of disturbed sediment. The near-surface zone will have increased turbidity in the vicinity of the mining vessel from the discharge of the air lift pumping system. This effluent will also have an increased content of nutrients characteristic of deep ocean water, so that the surface discharge will result in increased biological productivity locally. Whether it will produce net harmful or beneficial effects cannot be determined until actual operations can be monitored.

Present information regarding mining disturbances is derived from small-scale tests and laboratory studies and is inadequate to predict the nature and scale of potential pollution expected from full-scale mining. Estimates of both the surface and bottom ocean effects of nodule mining are highly conjectural at the present time, and the most meaningful insights can probably only be gained by careful study of natural phenomena analogous to mining disturbances.[13]

Moreover and with regard to the Area, the Convention contains a number of provisions whose purpose is to ensure that, *inter alia*, contractors take the necessary measures to protect the marine environment, that States and International Organizations co-operate to eliminate the effects of pollution or minimize its damage by jointly developing contingency plans for this purpose and that the Council within the scope of its mandate, issue orders for the suspension or adjustment of operations by a contractor if its activities cause or may cause serious harm to the marine environment.[14]

From the viewpoint of the potential investor in a land-based nodule processing plant and the concerned authorities in the potential host developing country, the relevant environmental perturbations are those directly associated with the plant.

The disposal of wastes resulting from processing ore is a major problem for the mineral industry and one the industry must be cognizant of. Depending on the extractive process, the rate of nodule mining and the form of waste disposal, wastes from a nodule-processing plant could accumulate at a rate of from 8 to 40 hectares (20 to 100 acres) per year (assuming a 12 m or 40 ft depth).[15]

Onshore environmental impacts are likely to occur in the port, during transportation, during processing and at the waste disposal facilities. As a result, the potential investor must consider the risk that the project will be delayed by challenges from environmental groups or protracted government decision-making caused by uncertainty; the risk that government regulations will be based on worst case assumptions and, therefore, be needlessly burdensome; and the risk that unanticipated effects detected after operations begin will require shutdowns and costly modifications.

A number of factors determine the production levels and process technologies to be utilized in first generation plants. The principal constraints on production levels will be:

(1) availability of capital for plants recovering copper, nickel and cobalt (the three metal option) and
(2) manganese market impact for plants recovering four metals (the four metal option).

The processing options described in the literature suggest that the state of the technological development of extractive metallurgy is such that first generation plants may employ 1 or all 5 of the following processing options.[16] For three metal plants, the following options will be used:

(1) gas reduction and ammoniacal leach;
(2) cuprion ammoniacal leach;
(3) high temperature and high-pressure sulphuric acid leach.

For four metal plants, the following options may be used:

(4) reduction and hydrochloric acid leach;
(5) smelting and sulphuric acid leach.

Table 10 summarizes the major outputs for the four metal option based on treatment processes determined to be the most likely to be used by the industry. In Table 10, "tailings" includes nodule rejects and/or slags, combustion ash and process solids, sludges and liquids. The problem to be addressed by the potential host developing country in accepting a nodule processing plant is whether or not the tailings can be disposed of within their existing regulations.[17]

Dames and Moore (1977) quantified and described waste streams for the gas reduction and ammoniacal leach option and for the pyrometallurgical process option. Their data included

Table 10. Products and major solid and liquid outputs for a pyrometallurgical (four metal) plant (annual basis)

Item	*Solids*	*Liquids*	*Units*	*Quantity*
PRODUCTS				
Nickel products			10^3 tons	40.6
Copper products			10^3 tons	35.5
Cobalt products			10^3 tons	5.7
Ferromanganese			10^3 tons	804.5
MAJOR TAILINGS, SOLIDS				
Nodule rejects (weight)[a]	2168			
(volume)[b]		676		
Scrubber sludges (weight)	290			
(volume)		108		
Ash (weight)	123			
(volume)		45		
Total (weight)	2851			
Total (volume)		829		
MAJOR TAILINGS, LIQUIDS				
Nodule rejects slurry (weight)	682			
(volume)		479		
Sludges slurry (weight)	378			
(volume)		367		
Ash slurry (weight)	955			
(volume)		955		
Total (weight)	2014			
Total (volume)		1800		

[a] Weight figures are given in 10^3 tons
[b] Volume figures are given in km^3
Source: After Dames and Moore Report, Vol. III, 4.3a, 8.4 (see Notes for details).

the residues and the impacts of the introduced reagents and fuels. The quantities and characteristics of the waste streams for the two processes are set forth in Table 11.[18]

Part of the solid wastes from the proposed three metal options consist of finely divided "tailings" or residues of nodules which have been chemically and physically altered and from which the desired metals have been extracted. This material would leave the plant in slurry form and would include small amounts of dissolved sea salts and trace elements.

Table 11. Characteristics of process waste streams[a] of a pyrometallurgical (four metal) plant (annual basis)

Content of waste streams	*Weight (10^3 tons)*	*Volume (km^3)*
Nodules, slagged and granulated (B, 1)	2168	676
Nodules, slag quench liquid (4, 5)	682	479
Sludges, solids, (A, 1, 2)	290	108
Sludges, liquid (4, 5)	378	367
Ash, solids (B, 1, 2)	123	45
Ash, liquids (B, 4, 5)	955	955
Total solids	2581	829
Total liquids	2014	1800
Total solids and liquids	4595	2629
Liquids impounded in rejects		1215
Net volume liquids for disposal or recycling		585

[a] Classification (after Dames and Moore Report, Vol. III, 4, 4.3a, 8.4)

Physical characteristics
A. Settleable sludges
B. Vitreous solids
C. Inert solids
D. Soluble solids

Chemical characteristics
1. Solid containing fixed toxic elements
2. Inert solids
3. Soluble solids with toxic elements
4. Solutions containing toxic elements
5. Solutions containing innocuous compounds

The physical characteristics of these sludges from the proposed three metal options would be very similar to tailings generated from the processing of nickeliferous laterites at the Moa Bay and Nicaro plants in Cuba. Their chemical composition, however, would differ, with residues from nodules processing consisting largely of manganese oxides and carbonates.[19]

The bulk of the solid wastes, however, from both the three and four metal options would differ significantly from the tailings from processing nickeliferous laterites. Residues from nodules processing will consist largely of granular slags. Such materials are known to be produced in the processing of ores of nickel and copper and it has been established that it is inert and stable over long periods. As a result, it is not expected to cause any significant problems.[20] A possible area of concern is in the reduction and hydrochloric acid leach option for which there is no analogous land-based operation. The wastes from this processing route will consist of roughly equal amounts of leached tailings accompanied by other process wastes and fused salts. The former would have properties akin to wastes from the three metal options but the latter would consist of dry bulk material. Since this material dissolves in water, it would have to be contained to prevent its dispersal into the natural environment.

Nodules also contain many components which appear on the list of hazardous or toxic substances in either their elemental form or in certain compounds. These total less than 0.5% of the nodules by dry weight. Table 12 identifies and groups the elements for each group. As nodules occur in nature, these constituents are chemically bound in a complex matrix and do not appear to be accessible to the environment from the actions of natural systems.[21,22] While harmless in their natural state, it is not currently known to what extent or in what manner these constituents might be transformed during processing.[23]

In order to obtain an idea of the environmental impact of tailings generated from a nodule processing facility and to determine whether or not such tailings could be disposed of within the United States without violating environmental regulations, the United States Bureau of Mines tested laboratory-generated tailings and slags for the five technically feasible first-generation nodule processing methods.[24]

In their tests, approximately 30–40 lbs. of tailings and about 125 lbs. of slags were produced for each process. Their overall extraction of the valuable metals, that is of cobalt, copper, nickel and/or manganese were generally greater than 90% for cobalt, copper and nickel, with lower values for the gas reduction and ammoniacal leach, and cuprion ammoniacal

Table 12. Major categories of elements in manganese nodules

Type	*Element*	*Percentage by dry weight of group in nodules*
Major and value metals	Manganese, Iron, Cobalt, Nickel, Copper, Zinc, Molybdenum	29% total
Innocuous non-minor elements	Sodium, Potassium, Magnesium, Calcium, Aluminium, Titanium, Silicon, Phosphorus, Sulphur, Chlorine	14% total
Known toxic elements, chemically bound	Barium, Lanthanum, Vanadium, Chromium, Silver, Cadmium, Thallium, Lead, Arsenic, Antimony	0.5%
Innocuous minor elements	Boron, Carbon, Scandium, Strontium, Yttrium, Zirconium, Niobium, Gallium, Tin, Bismuth	0.3% total 44% total
Oxygen as oxides and pore water		56% total
		100% total

Source: Dames and Moore and EIC Corp., 1977.

leach based processes (laboratory and pilot plant) for cobalt and copper.

In their tests, all tailings and slags were analysed for physical properties (except smelting and sulphuric acid leach), mineral content, elemental and ion content and leachability using the toxicity test of the United States Environmental Protection Agency, the shake extraction test of the American Society for Testing and Materials (ASTM) and the United States Environmental Protection Agency/Corps of Engineers' (EPA-COE) seawater elutriant test.[25–27]

The toxicity test is used to determine whether a waste is hazardous or not. The ASTM shake extraction test is an alternative to the Environmental Protection Agency toxicity test and was performed for comparison purposes. The United States Environmental Protection Agency/Corps of Engineers' (EPA-COE) seawater elutriant test was used to determine what element might leach in a seawater environment in the case of ocean disposal of nodule tailings.

For the ammonia-based processes, their tests revealed that the physical properties, mineral content and elemental and ion content were very similar with no significant variation in composition and physical properties. The high-temperature and high-pressure sulphuric acid leach process tailings physical properties were similar to the ammonia-based processes. The minerals Mn_3O_4 and MnO_2 were identified as minor (10 to 25 wt.%) constituents. Chemical analyses were not significantly different than the ammonia-based processes. In all of the three processes that do not recover manganese, the tailings, they determined, consisted mainly of iron oxide and hydroxides, manganese oxides and for the ammonia-based processes, $Mn\ Co_3$.

The reduction and hydrochloric acid leach process tailings were different from the ammonia-based and high-temperature, high-pressure sulphuric acid leach tailings, being less dense, of finer particle size and devoid of manganese. The primary constituents were iron oxides and hydroxides, quartz and feldspars. Chemically, the tailings were predominantly iron, silicon and aluminium. The levels of chromium and nickel observed in these tailings were considered high.

The slags from the smelting and sulphuric acid leach processes were of two types: the smelting slags and the ferromanganese slags. Each slag was analysed both mineralogically and chemically. The smelting slags were predominantly Mn_2SiO_4 and were used to produce ferromanganese. The slags from the ferromanganese tests were predominantly Ca_2SiO_4 and $Ca_2A\ SiO_7$.

The results obtained in their study showed that the tailings and slags produced in the laboratory and the pilot plant were non-hazardous as defined in the Environment Protection Agency's toxicity test and that none of the tailings and slags exceeded the maximum allowable limit for the eight metals (silver, arsenic, barium, cadmium, chromium, mercury, lead and selenium) in question. Furthermore, the results of the shake extraction test (ASTM) and the seawater elutriant test (EPA-COE) gave lower values than those obtained for common elements in the toxicity test.

Virtually every country has a scheme for requiring an environmental impact assessment by any party proposing something as large as a nodule processing plant. Such a requirement often tends to be statutory or executory and also tends to be informational only. The information provided by the assessment generally goes by law to the heads of departments charged with granting or denying other needed permits or licenses for such a processing plant.

Virtually every country also has a substantive code for air and water pollution control, usually complete with emission or effluent standards for certain pollutants. These can be expected to apply to particulate emissions from a processing plant as well as wastewater and tailings disposal into the marine environment.

In their survey of land and environmental laws and business/economic factors applicable to the siting of a manganese nodules processing plant in the Pacific basin/rim, Callies and Johnson (1985) state, *inter alia*, that

> Processing manganese nodules outside the United States is apparently sufficiently attractive to various consortia that several countries have been investigated for such purposes. Among the advantages claimed are proximity to known nodule deep sea beds and lower labour costs. However, those Pacific Rim/Basin countries with sufficient proximity advantages as well as sufficiently developed (or readily developable) skilled and semi-skilled labour forces and infrastructure (power, communications and the like) and port facilities to make such a site feasible also appear to have land use and environmental regulation/impact codes in various stages of development which clearly affect the siting of a manganese nodule processing plant, not only in terms of permit/permission processing time, but also in terms of costs of submission and permission exactions. It is, of

course, true that the cost of complying with various land use and environmental protection and impact laws in the United States is also expensive for the same reasons.[28]

Country Political Risk

One of the major concerns of both lenders and sponsors of a nodule processing facility is the stability of the political environment of the country selected, i.e. the political risk. Political risk as understood in this study will be in its broadest sense. It refers to those circumstances:

> (1) when discontinuities occur in the business environment, (2) when they are difficult to anticipate, and (3) they result from political change. To constitute a risk, these changes in the business environment must have the potential for significantly affecting the profit or other goals of a particular enterprise.[29]

In their study, Callies and Johnson (1985) state, *inter alia*, that:

> Political risk is defined as the probability that political events and conditions will have an impact on international business operations and profits (O'Leary and Coplin, 1981). Both political instability and government restrictions are two major subcategories of political risk.[30]

For a foreign investor, a political risk may arise from, *inter alia*, four main sources. Firstly, from changes in the regime, i.e. in the laws and regulations, under which the whole economy of the host country operates. Examples of significant changes that may affect the project's profitability are: introduction of a new tax system, import restrictions, new labour regulations, currency devaluation, environmental policies, transfer of technology provisions, etc. It is worth noting that some of these changes—for instance, a possible tax reduction—may have a favourable impact on the project's return. Banks consider that the way a government manages its domestic economic policy (policy itself, management quality and its political support) is the most important factor that should be taken into account when assessing political risk.[31] From the foreign investor's point of view, risks associated with changes in the economic regime of a host country are not perceived as significantly different from any similar risks he would face in his own country.[32] Secondly, the feasibility of a given project may be at stake if adverse political developments towards foreign investors eventually take place. These include developments such as restrictions on profit remittances (or any other kind of limitation or ban on the transfer of funds abroad), expropriation, nationalization, and forced disinvestment. Thirdly, political risk may arise from civil war, revolution, terrorism, etc. A secondary effect might be import restrictions imposed by other countries in response to political changes. Finally, another source of political risk may emerge from the possible changes that may be introduced in terms of the operating agreement that the foreign company has with the host country: for instance, increase in royalties, export restrictions on the project's output, increase in governmental control and participation in the project.

A manganese nodule processing plant will be within the jurisdiction of a given host country. Its activities will therefore be regulated by the local laws and the development agreement the investor has with the host government. Additionally, a candidate country for a processing plant not only has to fulfil its basic operational requirements (infrastructure, labour, power supply, etc.), but also has to present a favourable investment climate (positive attitude toward foreign capital, advantageous tax system, willingness to contribute to the project and healthy overall economic situation).[33] Since no specific site has yet been selected for a nodule processing facility by any of the consortia, the following analysis will focus on the possible issues to be addressed by both lenders and investors when assessing the political risks associated with the processing stage of a seabed mining venture.

Both lenders and investors will carefully analyse the political and economic characteristics of the countries where the processing facility might be located. For the sponsors of a manganese nodule venture, the expected return on investment might be estimated in the form of a probability distribution with its associated mean value and standard deviation. For the lenders, the best they can expect from a lending operation is that payment of interest and principal is done according to an agreed schedule.[34] Moreover, it has often been argued that the risk for the lender is limited to the amount of loan he advances to a project while the sponsor has the whole project at stake.

The different perceptions of political risk are dependent on the economic agent to whom the loan is made. On the one hand, if a loan is channeled to a project, and its servicing is subject to the future returns of that project, one can expect that both lenders and borrowers may perceive political risk the same way. On the other hand, if a loan is made to a host government which is expected to provide the project with the required infrastructure, a lender may perceive political risk differently from the investor. In this case, banks will look at the financial ability of the country to honour its debts as well as to its political willingness to do so. Depending on the country, a perception of higher risks may merely imply higher financing costs to the borrower. For some other countries, however, the availability of financial resources, at whatever the price, may decrease as a consequence of perceived higher risks.[35]

During the political risk assessment process, lenders will address issues such as: (1) the stability of the government involved in the project; (2) the perception of both the sponsoring and the host states of the importance of the project (in case a project is sponsored by one country and is located in another one); (3) the type of equity or other financial support both states are likely to contribute to the investment; (4) the likely stability of the labour force promised within a particular state; (5) the state involvement in project management and the competence of individuals selected; (6) legislative (including tax) benefits allowed by the host state; and (7) the stability of the currencies in question and assurances of its convertibility.

Evidence seems to suggest that political risk assessment is done in a much more systematic way by banks than by most direct foreign investors. Several methods may be used for political risk assessment. Grosse and Stack argue that the most used techniques are: (1) the checklist method which consists of a numerical scheme of a country classification where a check list of relevant factors are scored on some scale; (2) scenario development which involves the forecast of economic and political conditions of the country concerned under different sets of assumptions; and (3) Delphi technique in which a group of experts is selected to answer a questionnaire on the assessment and forecast of certain outcomes in a target country. The experts answer the questionnaire separately and later their opinions are put together and given back to them for revision.[36]

It has been often argued that the basic problem of political risk assessment methods is that they are very subjective, i.e. they rely on subjective knowledge or perception one has on the country in question. Although such information may provide the lender or investor with an idea of the volatility of the political and economic environment, the probability of the information "being able to successfully predict all major risks especially along with their timing, is small".[37] Lenders' and investors' attitude toward political risk is dependent on its perceived magnitude. The magnitude of the risk can be translated into a decision of not making the investment on loan, the use of insurance or hedging mechanisms or the requirement of a higher spread or a higher rate of return on the investment.

DEVELOPING COUNTRY OBJECTIVES

Many countries would want to have a nodule processing plant located within them. This plant, however, would have to meet many objectives related to mineral projects. These objectives would often be pursued through joint ventures with foreign investors, through the operations of state mineral enterprises, or through concession agreements that permit others to exploit a country's mineral deposits. Five of the principal objectives are discussed below and these are:

(1) balance of payments and foreign exchange earnings;
(2) revenue objectives;
(3) transfer of technology;
(4) employment and training; and
(5) forward and backward linkages with the national economy.

Balance of Payments and Foreign Exchange Earnings

A key objective of most countries is to gain as much foreign exchange as possible in efforts to pay for imports. The key purpose for exporting is to gain foreign exchange. Foreign exchange is also gained when an outsider brings in foreign capital for development purposes. There is a tendency to focus on the balance of payment accounts rather than the foreign exchange that is necessary for a country to survive. The balance of payments is a system of accounts designed to show how a nation finances its international activities. The focal point of the balance of payments is the nation's external liquidity. This measures the country's ability to acquire the goods and services it needs to import and to pay for them. The accounts are set up to summarize all economic transactions between the residents of the reporting country and the residents of other countries during a given period of time, normally a year or a quarter. Although the balance of payments' primary purpose is to answer questions pertaining to a country's international financial position, it also serves as a framework for analysing and interpreting a wide range of problems dealing with the country's economic and business life.

Nodule processing plants would require large amounts of capital for development purposes. A developing country that is successful in being selected to host such a plant would obtain a very large and positive inflow of funds during the construction phase. Additionally, the inflow of foreign exchange from sale of metals once operations commence should be more than adequate to service the subsequent outflow to pay back principal, interest and dividends.

The host country can also share in the foreign exchange

earnings through taxation or equity participation in the nodule processing venture. However, there will be periods of low metal prices when foreign exchange earnings could suffer considerably and the operation could be adversely affected, especially if a high proportion of the production cost is represented by foreign exchange input. There are two types of cases which illustrate this point.

First, there are countries like Zambia and Zaire which have a high proportion of their production cost represented by foreign exchange inputs. These countries employ expensive expatriate labour; they have long supply lines over land and sea on which the goods have to be paid for in foreign currency; they have to import much, if not all, of the plant, equipment and supplies required to produce copper; and they have to meet heavy interest charges on foreign loans. Therefore, only a part of the revenue they earn from each ton of copper on international markets is a net gain in foreign exchange. The amount of this gain depends, of course, on the price of copper.

Second, there are countries like Chile and Peru which have much shorter distances to suppliers and to their markets; moreover, they do not employ such a large proportion of expatriate labour, and they have internal sources of supply for some of the goods required for refined copper production. In countries such as these, the foreign exchange input in the cost of producing copper may be small, so that a substantial part of the total revenue from its sale represents net earnings in foreign exchange.

Revenue Objectives

For many developing countries, the revenue derived from mineral operations provides a substantial portion of government income and is essential in national economic planning. The situation may be particularly hazardous for small countries which rely solely on a single mineral operation for a large percentage of the country's foreign exchange earnings. For example, in 1981 approximately 71% of Mauritania's total exports could be attributed to iron ore exports. If the iron ore business were to fail, then this source of government revenue would also fail.

Numerous governments have found that mineral production represents a highly variable and unpredictable source of income. This is especially true where government income is related to the profitability of the venture either through government equity participation or by taxation. Consequently, the government will experience planning problems because of fluctuating and unpredictable revenues. Some governments such as the provincial government in Alberta, Canada, have established mineral resource stabilization funds into which mineral revenues are paid and from which a relatively steady and predictable flow of funds may come into the government budget.

Transfer of Technology

From the point of view of developing countries, transfer and practical application of technological knowledge is considered as one of the essential benefits of foreign investment. Gaining know-how offers promise of narrowing the gap in levels of industrialization and promotes socio-economic development on a long-term basis. In particular, transfer of technology is aimed at the following goals: growth of employment and purchasing power; growth of output and more effective production; a more even distribution of income; and fulfillment of the basic needs of the population.[38]

In practice, technology and know-how are traded as commodities. In cases of different technological levels between the seller and the recipient, transfer is accomplished by some form of contractual arrangement. Technology may be transferred in many ways including:

(1) flows of books, journals and other scientific and technical publications;
(2) the movement of people including the inflow of technical, scientific and management personnel, and the flow of

people on foreign educational, training and observation assignments;

(3) direct foreign investment accompanied by equipment and personnel;

(4) the direct importation of machinery and equipment;

(5) licensing, patents and know-how agreement; and

(6) technical assistance programmes of governments, on either a bilateral or a multilateral basis.

Special to manganese nodule mining, transfer of technology comprises the following objectives: development of suitable marine technology; collection and dissemination of data and knowledge on marine technology; and harnessing human resources by means of technology transfer. The objectives may be achieved, in part, by training and education of nationals of all countries and programmes for technical co-operation at the international, regional and bilateral levels.

Employment and Training of Nationals

In most countries, full employment is of particular concern. Foreigners are often invited to invest and provide jobs at the same time, countries seek to use the foreigners as a source of know-how to train and increase the level of skills available to the nation.

Nodule processing plants would require perhaps less than 1000 employees and that quantity by itself is not sufficiently high to ensure a rush of countries to demand processing plants. Training may be of greater concern.

In general, when a foreign firm enters a country, it is common for government agreements with foreign investors to require the firm to give preference to the hiring of nationals of the host country, to meet minimum quantity goals of local employment and to provide training. Preference can often be accommodated and the quantity of employees is easy to determine. However, training programmes are difficult to control. Training is both time and knowledge related. Few governments can adequately check to see if the programmes

Table 13. Progress in turning jobs over to nationals

Training skill category	*After 3 years (%)*	*After 5 years (%)*	*After 8 years (%)*
(1) Unskilled	100	100	100
(2) Skilled	50	75	100
(3) Clerical	50	75	90
(4) Technical/supervisory	50	75	85
(5) Management/professional	50	75	85

The percentages indicate the level of each employment classification that should be held by nationals within the periods stated beginning with the commencement of the operating period.

meet a minimum quality standard and the results of the training programme may not be apparent for many years. This is especially true of managerial training programmes.

Most governments require that locals be hired for certain posts in the expectation that this gives the company incentive to train personnel adequately. A common requirement for investors and one which has worked well on occasion, is to have a higher percentage of local workers for the jobs requiring less skills and providing a timetable that requires progress towards turning managerial jobs over to trained and qualified local workers. The burden of justifying exceptions is thus placed on the company. Table 13 outlines what a typical timetable might be.

Provisions that require the employment of nationals work best when they require that the company submit to the government (1) its reasons for any request for exceptions, and (2) documentation demonstrating the adequacy of its past training programmes which were designed to accomplish the goals of providing the required skills to local workers.

Although the discussion above focused on the employment opportunities offered by the plant itself, the socio-economic impact of such a plant is even more profound. A social impact analysis explores how a proposed new industry or land development can affect the way people live and interact with one another. An economic impact analysis explores how a proposed new industry or land development can affect the economy of a community. Thus, a socio-economic impact analysis is an analysis of the impact on the daily life and business transactions of the members of a community.

In undertaking an initial assessment of the potential economic impact of establishing a manganese nodule processing plant in Hawaii, the Department of Planning and Economic Development of the State of Hawaii concluded, *inter alia*, that:

(1) during the construction phase of the plant, the increased output of the county of Hawaii would range from US$ 119 million to US$ 221 million, with an increase in personal income ranging from US$ 38 million to US$ 55 million, and an increase in annual employment ranging from 1572 to 2684;
(2) in the operation phase, the increase in output of the county would range from US$ 535 million to US$ 836 million, personal income from US$ 29 million to US$ 41 million and employment from 889 to 1357.

In arriving at these estimates, the Department took into account the allocation of operating costs to sectors. The sectors they distinguished included: *households* for labour directly associated with the processing plant and labour costs for the transport crew; *maintenance and repair services* to the consortium-owned transport ships through established firms in Honolulu; the *wholesale trade sector* through wholesalers on the island which is expected to supply all the process chemicals which are required in annual quantities less than US$ 100 000; and *banking, finance, and value-added.*

Forward and Backward Linkages

A major concern of a potential developing host country will be the contribution to its economy of a nodule processing plant and related facilities. One of the economic factors is the backward and forward linkages resulting from the industry. Backward linkage relates to the development of ancillary facilities, supplies and labour purchased from other upstream industries. Forward linkage relates to the downstream industries and measures the amount of output of a particular industry that produces intermediate inputs from the products of the processing plant for other industries within the economy.

For primary industries, backward linkages do not usually create great impact because these industries usually exist as inputs to other industries and it is from those forward linkages that their largest contribution to the economy is derived. For example, a prime illustration of large forward linkage is the forestry industry which sells almost all of its output to the manufacturing sector. Conversely, for the metal mining sector, there is very little forward linkage given the paucity of processing facilities in the developing countries (see Table 1), as most of the output is usually exported to other countries in the form of ores and concentrates.

A nodule processing plant, however, presents opportunities for both forward and backward linkages. On the one hand, opportunities exist for the country in question particularly if it has the raw material base to provide local sources of material inputs to the industrial concern (backward linkages) and, on the other, opportunities are created for the production of semi-manufactures and manufactures (forward linkages). Paramount in the nodule processing industry is the need for power, large amounts of electrical energy, in the order of from 300 to 400 megawatts, depending upon the size and type of plant, i.e. the tons of nodules to be processed and the type of process selected. Supply possibilities for this significant backward linkage could be from plants providing hydro, coal, oil, gas, nuclear and possibly geothermal power.

The International Energy Agency recently published a study comparing the relative competitiveness of coal, oil and nuclear fuel for electricity generation.[40] The study concluded that depending on operating rates and the sulphur content of the coal, thermal coal is more economical to use than oil when coal prices are below US$ 65–86 per ton of coal equivalent (tce) (in 1978 terms). In fact, in terms of heat value, coal costs are about half as much as oil.

The beginning of the process for forward linkages is at the smelting and refining stages and the process concludes in the production of automobiles, pipes, building materials and a variety of other products. This stage, for a large number of developing countries, is only attainable with the siting of the processing plant in the country concerned.

Summary

Developing countries wishing to have a nodule processing plant located within them would do so for a number of reasons. A key objective of most countries would be to gain as much foreign exchange as possible. Nodule processing plants would require large amounts of capital for development purposes. Developing countries can, therefore, obtain a very large and positive inflow of funds during the development and construction stages of a processing plant.

In sharing in the foreign exchange earnings obtained through taxation or equity participation in the nodule processing venture, developing countries will seek, *inter alia*, that nationals be trained (to decrease the costs of employing expatriate personnel who have to be paid in foreign exchange), develop internal raw materials for supplying (to decrease the outflow of foreign exchange), and provide some of the services required by the venture and increase the technological knowledge of their nationals.

Finally, in developing countries certain developmental objectives could not be fulfilled because, for reasons of lack of market size, some of their resources had never been developed to their full potential. Developing countries, thus, might view such a processing facility as a central node as a result of which opportunities might open up for their resources to be fully developed and for their developmental objectives to be realized.

NOTES

[1] Article 156 of the United Nations Convention on the Law of the Sea (Establishment of the Authority).

[2] Article 144 of the United Nations Convention on the Law of the Sea (Transfer of Technology):

"1. The Authority shall take measures in accordance with this Convention:
   (a) to acquire technology and scientific knowledge relating to activities in the Area, and
   (b) to promote and encourage the transfer to developing States of such technology and scientific knowledge so that all State Parties benefit therefrom.

"2. To this end, the Authority and States Parties shall co-operate in promoting the transfer of technology and scientific knowledge relating to activities in the Area so that the Enterprise and all States Parties may benefit therefrom. In particular they shall initiate and promote:
   (a) programmes for the transfer of technology to the Enterprise and to developing States with regard to activities in the Area, including, *inter alia*, facilitating the access of the Enterprise and of developing States to the relevant technology, under fair and reasonable terms and conditions;
   (b) measures directed towards the advancement of the technology of the Enterprise and the domestic technology of developing States, particularly by providing opportunities to personnel from the Enterprise and from developing States for training in marine science and technology and for their full participation in activities in the Area."

For the provisions related to financing the Enterprise, see Annex IV of the Convention.

[3] See last paragraph of Chapter 3.

[4] Preparatory Commission for the International Sea-Bed Authority and for the International Tribunal for the Law of the Sea, 1987. Draft regulations on prospecting, exploration and exploitation of polymetallic nodules in the area. (Draft regulations on financial incentives) (Document *LOS/PCN/SCN.3/WP.6/Add.3,* 11 March 1987).

[5] D. L. Callies and Johnson, C. J. 1985. "A survey of land and environmental laws and business/economic factors applicable to the siting of a manganese nodules processing plant in the Pacific basin/rim: Australia, Canada, Colombia, Ecuador, Fiji and the Philippines." University of Hawaii Sea Grant College Program, Grant No. NA81AA-D-00070 Sea Grant Report UN1H1-SEAGRANT-CR-85-02, April 1985.

[6] (a) US Department of Commerce, 1979. "Gulf Coast Manganese Nodule Processing Plant Location Criteria."
(b) Flipse, J. E., Andrews, B. V. and Brown, F. C. 1983. "The economic viability of a four-metal deep ocean mining venture." Washington, D.C. National Oceanic and Atmospheric Administration.
[7] (a) Pasho, D. W. and King, D. E. C. 1980. "Processing sector cost estimates for the Ocean Management Inc. manganese nodule processing facility." Mineral Policy Sector Internal report MRI 82/7 Canada Department of Energy, Mines and Resources, August 1980.
(b) Pasho, D. W. and King, D. E. C. 1980. "Processing sector cost estimates for the Kennecott joint venture manganese nodule processing facility." Mineral Policy Sector Internal report MRI 82/5 Canada Department of Energy, Mines and Resources, August 1980.
[8] Hawaii Department of Planning and Economic Development. 1981. *Feasibility and potential impact of manganese nodule processing in the Puna and Kohala districts of Hawaii.* Prepared by the Department of Planning and Economic Development, State of Hawaii and the US Office of Ocean Minerals and Energy, Honolulu: Department of Planning and Economic Development, State of Hawaii.
[9] US Department of Commerce, op. cit.
[10] Hawaii Department of Planning and Economic Development, op. cit.
[11] Most of the material in the following sub-sections was obtained from Deep Sea-Bed Mining—Final Programmatic Environmental Impact Statement. Volume I. US Department of Commerce. National Oceanic and Atmospheric Administration. Office of Ocean Minerals and Energy. September 1981.
[12] See Appendix 1 for details on coal availability in selected regions of the world.
[13] IMCO/FAO/UNESCO/WMO/WHO/IAEA/UN. Joint Group of Experts on the Scientific Aspects of Marine Pollution (GESAMP) 1977. *Scientific Aspects of Pollution Arising from the Exploration and Exploitation of the Sea-Bed.* GESAMP Reports and Studies, No. 7.
[14] United Nations Convention on the Law of the Sea:
(i) Article 145 of the Convention (Protection of the Marine Environment);
(ii) Article 162 of the Convention (Powers and functions of the Council);
(iii) Article 199 of the Convention (Contingency plans against pollution).
[15] Deep Seabed Mining—Final Programmatic Environmental Impact Statement. Vol. 1, op. cit.

[16] United Nations, 1986. *Analysis of Processing Technology for Manganese Nodules*, op. cit.

[17] Hawaii Department of Planning and Economic Development, op. cit.

[18] Dames and Moore and EIC Corporation. 1977. *Description of Manganese Nodule Processing Activities for Environment Studies.* Volume II, *Transportation and Waste Disposal Systems.* (Contract 6-35331.) United States Department of Commerce—NOAA, Office of Marine Minerals, Rockville, MD. NTISPB274912.

[19] United Nations, 1986, *Analysis of Processing Technology for Manganese Nodules*, op. cit.

[20] Deep Sea-bed Mining—Final Programmatic Environmental Impact Statement. Vol. 1, op. cit.

[21] United Nations, 1986. *Analysis of Processing Technology for Manganese Nodules*, op. cit.

[22] Deep Sea-bed Mining—Final Programmatic Environmental Impact Statement. Vol. 1, op. cit.

[23] Ibid.

[24] Haynes, B. W., Barron, D. C., Kramer, G. W., Maeda, R. and Magyar, M. J. 1985. *Laboratory processing and characterization of waste materials from manganese nodules.* US Bureau of Mines, Report of Investigations RI 8938, 16 pp.

[25] Federal Register. Parts II–IX, *Environmental Protection Agency: Hazardous Waste and Consolidated Permit Regulations.* Vol. 115, No. 98, May 19, Book 2, pp. 33063–33285; 110-CFR, Parts 260–265.

[26] American Society for Testing and Materials. 1982. *Standard Test Method for Shake Extraction of Solid Waste with Water.* D3987 in 1982 Annual Book of ASTM Standards: Part 31, Water. Philadelphia, PA, pp. 1423–1427.

[27] US Environmental Protection Agency/Corps of Engineers. 1977. *Ecological Evaluation of Proposed Discharge of Dredged Material into Ocean Waters.* Environmental Effects Laboratory, Waterways Experiment Station, US Army Corps of Engineers, Vicksburg, MS, July 1977, 103 pp.

[28] Callies, D. L. and Johnson, C. J. 1985, op. cit.

[29] Robock Stefan cited by Grosse, R. and Stack, J. "Non economic risk evaluation in multinational banks", *Management International Review*, Vol. 24, No. 1, 1984.

[30] Callies, D. L. and Johnson, C. J., 1985, op. cit.

[31] De Vries, R. 1983. "Country Risk, a Banker's View", in Herring, R. J., *Managing International Risk.* Cambridge University Press, New York.

[32] Reid, J. T., "Managing Country Risk for a Manufacturing

Corporation" in Herring, R. J., *Managing International Risk*. Cambridge University Press, New York.

[33] Callies, D. L. and Johnson, C. J., 1985, op. cit.

[34] Grosse, R. and Stack, J., op. cit.

[35] World Bank, *Private Bank Lending to Developing Countries*. Washington, D.C., World Bank Staff Working Paper No. 482, 1983.

[36] Grosse, R. and Stack, J., op. cit.

[37] Reid, J. T., op. cit., p. 186.

[38] Technology includes the specialized equipment and technical know-how, including designs, operating instructions, training and the technical advice and assistance necessary to assemble, maintain and operate a viable system.

[39] Hawaii Department of Planning and Economic Development, op. cit.

[40] *Steam Coal Prospects to 2000*, International Atomic Energy Agency, 1978.

Chapter 5

PROCEDURE FOR EVALUATING A LAND-BASED PROCESSING SITE

Once a decision has been made as to the type of nodule producing plant that is to be built, for the investor the site selection process begins. Apart from the essentially economic and technical component of the evaluation which constitutes the subject of this chapter, the potential investor would also undertake analyses of a country's political risk. In most cases, the latter analyses would be undertaken only for those countries that qualify when a weighted criteria method is utilized to establish that they meet the minimum economic and technical requirements.

In the weighted criteria method, once the requirements of a new nodule processing facility are firmly decided (i.e. financial requirements, land, manpower, utilities and services) they can be superimposed on a number of possible locations and compared against the same set of criteria. By assigning numerical values to each criterion to show its relative importance for the particular project, the potential sites can be weighed and compared to find the preferrable ones.

The two most important factors to consider when using a weighted criteria method are:

(1) the initial maximum values assigned to the criteria; and,
(2) recognition that the site selection process is a dynamic, not static one.

The values assigned to the criteria may have to be altered from time to time to account for changing conditions in the country. A characteristic of many developing countries is that changes

Table 14. Weighted criteria method to select a nodule processing site

Criteria	*Maximum value assigned*[a]	*A*	*B*	*C*	*D*	*E*	*F*	*G*
(1) Distance to port	107	107	214	167	214	107	107	214
(2) Port facilities	30	40	30	35	30	35	40	30
(3) Electrical power/fuel	385	390	450	500	420	400	470	520
(4) Access to markets	30	35	35	35	40	30	45	35
(5) Land	30	30	30	30	30	30	30	30
(6) Water	75	75	75	80	85	80	90	85
(7) Transportation at port	15	20	15	25	20	20	20	25
(8) Local raw material supplies	100	150	200	250	100	125	150	225
(9) Nickel laterite ore	42	250	60	100	250	250	200	150
(10) Labour availability	136	250	190	250	272	200	200	170
(11) Waste disposal	35	35	35	105	70	35	105	105
(12) Living conditions	15	30	15	15	30	45	15	30
(13) Incentives offered by ISA	100	N/A	N/A	N/A	N/A	N/A	N/A	N/A
(14) Incentives offered by host	100	N/A	N/A	N/A	N/A	N/A	N/A	N/A
(15) Prestige: effect on company reputation	50	N/A	N/A	N/A	N/A	N/A	N/A	N/A
Applicable sub-total	1000	1412	1349	1592	1561	1357	1472	1619
Total	1250							

[a] The lower the value scored on a given criterion, the more it fits into project economics.
Source: Adapted from E. S. Groo, "Choosing foreign locations: one company's experience", *Colombia Journal of World Business*. September–October 1971, p. 77.

are taking place within them at a fast pace; their internal markets are expanding and changing. Changes in the availability of power, raw material resources, and infrastructure could occur at a very fast rate and may be expected in some cases. Consequently, static analysis may provide misleading answers to the site selection questions.

Table 14 is an illustration of a weighted criteria method to select a processing site. The weights assigned to each of the criteria were obtained by comparing the annual operating cost parameters of relevance to any potential site as presented in Table 2 (Selected cost models for an integrated nodule venture). In this particular example, the variables of relevance for a four metal processing plant using the data provided in the cost models (Table 2), on an annual basis, are used. These variables are:

(i) annual transport costs from mine site to port;
(ii) annual energy costs;
(iii) annual labour costs; and
(iv) the variable miscellaneous costs that include costs for water, land, waste disposal, raw material inputs, etc.

Mining costs would be irrelevant to the decision as they would be the same irrespective of the location of the processing plant.

Table 15 provides a summary of how the weights for each variable were obtained. The percentages of the total annual costs for transporting nodules from the mine site, the energy costs associated with their processing and the costs for labour, and other miscellaneous costs from each of the four cost models for a four metal venture were averaged to obtain the comparative cost contribution of each of them to the operation on an annual basis (column 6 of Table 15). On the basis of the data provided in the cost models, the weights to be assigned the individual elements, i.e. transportation from mine site to processing site, energy costs associated with processing, labour and variable miscellaneous costs were determined to be 10.7, 38.5, 13.6 and 37.2 respectively.

In Table 14, the weights are further subdivided among the

Table 15. Assignment of weights

Variable	Annual operating cost (million US dollars)				Sub-totals	Per cent of total
	Hillman (1)	Hillman (2)	Ingham (Aus.)	Ingham (Ent.)		
	(1)	(2)	(3)	(4)	(5)	(6)
Transportation	13	11	5	5	34	10.7
Energy	27	28	34	33	122	38.5
Labour	10	10	12	11	43	13.6
Miscellaneous	28	28	30	32	118	37.2
Total	78	77	81	81	317	100.0

variable miscellaneous costs items (items 2, 4 to 9, 11 and 12). Items 13 to 15 on the other hand could not be determined at the present stage.

In this example sites B and E scored best, and would be deemed worthy of country political risk analysis. The final selection of a site would take place after this analysis by undertaking engineering, economic and legal feasibility studies of the most attractive site derived from this "short" list and comparing them. However, the number of such expensive and time-consuming feasibility investigations would be limited to the top contenders—all acceptable sites—as measured by the weighted criteria method.

Chapter 6

POTENTIAL SITES FOR FIRST GENERATION NODULE PROCESSING FACILITIES

With some of the noduliferous areas in the Central North-eastern Pacific Ocean as well as the Central Indian Ocean basin having been allocated to pioneer investors (Japan, France, USSR and India) and designated as reserved areas for the conduct of activities by the Authority in the Area through the Enterprise or in association with developing States, the first in a series of steps that could result in nodule mining and processing has been taken. For pioneer investors, their certificates of registration accord them the exclusive right to carry out pioneer activities in the allocated areas.[1] If, through these activities, registered pioneer investors should decide to undertake exploration and exploitation of polymetallic nodules in their allocated areas, within six months of the entry into force of the Convention, they are to apply to the Authority for approval of a plan of work.

Sites for nodule processing facilities will therefore not be selected until the results of pioneer activities have been analysed and a decision taken as to whether or not to apply for approval of a plan of work for exploration and exploitation.

Although deposits of manganese nodules occur in every major ocean, according to the trade literature, the area of greatest commercial interest lies within the Central North-eastern Pacific Ocean. The greatest potential within this belt is said to lie in a narrow band defined by longitudes 20°N and 40°N and latitudes 120°W and 180°W. Mine sites in this region would be chosen not only because of their technical merit but because of their proximity to markets for processing and sale.

The second area that has generated significant interest is within the Central Indian Ocean basin wherein a pioneer area has been allocated to the Government of India and another designated as a reserved area for the conduct of activities by the Authority.

First generation onshore processing plants would likely be located along the Pacific or Gulf coasts in the case of those deposits in the Central Northeastern Pacific or along the coastal states that border the Indian Ocean basin in the case of those deposits in the latter area, to minimize the cost of transporting raw nodules.

Although perceptions and judgements of country political and economic risk will affect the decision as to where to locate a plant, the initial consideration will be based on project economics including considerations of how to lower project risks since mining sector risks are considered to be significantly higher than processing sector risks.

The crucial economic factors in selecting a site for an onshore manganese nodule processing plant will be:

(1) transport costs from mine site to the plant;
(2) energy input costs for the operation;
(3) the cost of raw materials for the plant;
(4) the cost of environmental regulations at the site.

In respect of lowering project risk, factors that will be considered crucial will be the availability of lateritic nickel ore so that if nodules are unavailable on a continuous basis and at specified quantities, alternative feed material is available.

In the case of deposits within the Central Northeastern Pacific ocean and in respect of transport considerations alone, onshore processing plants would appear to be favoured:

(1) on the North American West Coast and the Hawaiian Islands in particular, and
(2) in the developing countries that are located on the West Coast of Central America.

If, for example, a Central American State were interested in competing for such a plant, using as input ore mined from one of the pioneer areas or an area designated as a reserved area, once that decision is taken, to overcome the disadvantage introduced purely by its distance from the mine site in comparison with Hawaii, it would have to be favourably disposed to offering a significant reduction in one or more of the remaining variable costs, such as energy and/or such raw materials as suitable lateritic nickel ore.

In the case of the Central Indian Ocean basin, the choice from a transport cost factor standpoint is significantly larger. In fact, ports in India, Malaysia, Sri Lanka, Indonesia and the Malagasy Republic are all competitive. In this particular case, other considerations could loom larger.

With regard to the pioneer area allocated to India, it would appear that the plant to process ore originating from this area would be located in India. As previously noted, India has developed its coal resources and continues to do so. India is also well endowed with skilled personnel who can be usefully engaged in processing activities.

Indonesia has a number of very good attributes to recommend it as a site for a processing facility for seabed manganese nodules. Aside from its competitive distance from one of the sites in the Central Indian Ocean basin, Indonesia is famous for low-sulphur oil and has significant resources of coal. Moreover, the state of development of its nickel laterite industry is well advanced with a nickel matte producing operation in Sulawesi and a ferronickel plant at Pomalaa.

NOTES

[1] Pioneer activities as defined by resolution II of the Third United Nations Conference on the Law of the Sea means "undertakings, commitments of financial and other assets, investigations, findings, research, engineering development and other activities relevant to the identification, discovery, and systematic analysis and evaluation of polymetallic nodules and to the determination of the technical and economic feasibility of exploitation".

Chapter 7

CONCLUSIONS

Of the two major components of a seabed manganese nodule venture, the processing component is the least risky. With the pioneer investors in deep sea mining having been allocated areas by the Preparatory Commission for the International Sea-Bed Authority and for the International Tribunal for the Law of the Sea, the nascent industry, it may be said, is in the process of being established. Once pioneer activities lead to a decision to undertake exploitation of polymetallic nodules, the quest for sites for a processing plant will take place in earnest.

The first generation of manganese nodule processing facilities are expected to be land-based. They are also expected to be highly energy intensive. Current economic studies of possible processing sites indicate that processing costs will also be sensitive to transport costs. These studies also suggest that the most risky aspect of the project will be the availability, on a continuous basis, of nodules.

For a site to qualify to be considered for the location of a manganese nodule processing facility, it will have to satisfy minimum economic and technical criteria that will be determined by the potential investor in respect of, *inter alia*, its infrastructure characteristics, distance from mine site, environmental and land-use laws, energy availability, and complementary inputs to the operation.

From the viewpoint of reducing economic risk, the site's potential would be boosted if it were within an acceptable distance from a nickel laterite orebody. Finally the investor will superimpose economic and country political risk criteria on those sites that qualify.

From a potential host developing country's viewpoint, a manganese nodule processing plant will mean large inflows of capital into the country during its construction and large revenues during operation. Although the facility itself will not be a large employer of labour, existing and new enterprises created to service this facility will bring additional employment and revenues.

For those countries interested in hosting such a facility, it is recommended that systematic efforts at preparing the requisite infrastructure and available raw material sources within the country be undertaken with a view to selling the attributes that the site has to potential investors.

Appendix

ENERGY AS A SPECIAL ASPECT OF SITE SELECTION

The energy crisis, which began late in 1973 and touched to some degree all countries, intensified interest in the future prospects for the development of world energy, particularly the contribution of coal. Although for at least a decade and a half the majority of experts in the energy field alleged that the coal industry had no future in an era of scientific and technological revolution, the situation has radically changed. The prospects of a resurgence in coal use have been enhanced by the increasing cost of other sources of energy, the delays in nuclear power programmes and the anticipated depletion of oil and natural gas supplies.

Coal constitutes a very important energy potential. World coal reserves which are exploitable with present technologies and at current prices are nearly five times larger than known oil reserves. Geological coal resources are several times more extensive. While coal is found in many places, 97% of presently known coal deposits are concentrated in developed and centrally planned economies. But there has been little coal exploration in developing countries and their basic geological data are inadequate. Comprehensive geological surveys and coal exploration programmes in developing countries may well increase their known coal reserves substantially. Many coal bodies need additional drilling and feasibility work before investment decisions can be made. Many developing countries also lack the infrastructure to utilize coal deposits that would be exploitable in industrial nations. The risks and high capital expenditures involved in coal mine development and the uncertainties of marketing opportunities have also meant that few coal deposits in developing countries have appeared attractive to investors. A coal deposit whose output could be utilized in a thermal power plant for a nodule processing facility could eliminate the uncertainty of marketing and, *ceteris paribus*, render a deposit potentially attractive to investors.

The conversion of coal to energy is based on well known and proven technology. Coal prices, however, vary, depending largely on the cost of production. In some countries, where production costs of coal exceed the price that it could normally receive in the market, government subsidy or market intervention occurs to maintain the viability of the domestic coal industry. This is the situation in Germany, Japan, Spain and the United Kingdom. In Australia, the United States and Canada, coal prices are more competitive, reflecting advantageous production conditions. Largely because of inland transport and handling cost savings, prices for domestic coal in these countries can be lower than those for export. Export prices are determined by commercial negotiation. As trade in coal increases, world price levels are likely to emerge. These should be essentially determined by the supply of and demand for coal and, given the overall availability and geographical distribution of potential coal supplies, should remain considerably below the world oil price on a thermal equivalent basis.[1]

Despite the fact that in terms of heat value alone coal costs are about half as much as oil, coal's competitive position relative to other energy sources remains ambiguous. The attractiveness of coal compared to other fuels depends not only on relative prices and calorific values but also on availability and reliability of supply, government energy policies, the satisfactory solution of technical and environmental problems that presently limit coal use and last, but not least, coal handling properties. Coal is difficult and expensive to distribute, particularly to small users and the need to control dust, dispose of ash and to provide adequate storage as well as to burn it in an environmentally acceptable manner makes it inconvenient to use. Nevertheless, the prospects for coal development have been further enhanced by the increasing cost, perceived risks and delays in nuclear power programmes and the prospect that limited oil and natural gas supplies will result in higher energy prices in the future.

Coal use also depends on the characteristics of interfuel substitution. International studies on fuel prices and fuel shares of individual countries by Griffin and Gregory (1976) and by Pindyck (1979) confirm that there is a high degree of substitutability between fuels in thermoelectric generation.[2,3] In these studies, fuel oil was shown to be almost as closely substitutable with coal as with gas. A study on coal substitutability by the OECD (1978) concentrated more on comparisons of electricity generating costs.[4] The major result of the study is that new coal-fired power stations could produce electricity at a lower cost than oil-fired power stations with similar environmental protection devices.

The fact that coal exploration in developing countries has been very limited and that resource data are incomplete is well known. Thus, any published figures have to be accepted as representing minimal quantities. In its 1979 study, the World Bank showed that about 50 developing countries had known coal resources while 20 had known recoverable coal reserves. Tables 16, 17 and 18 summarize the available information on developing country coal reserves and resources. India is shown to have the largest recoverable reserve base.[5] India, together with Indonesia, Brazil, Botswana and Swaziland account for 77% of the coal resources of the developing countries and 62% of their economically recoverable reserves.

In several developing countries, coal planning has actually taken place. The experience of India, Indonesia, and Colombia are cases in point. Among developing countries, India has long depended on coal to meet its energy requirements. This has been possible because of India's abundant reserves of coal relative to oil. Coal has played a major role in the development of India's commercial energy and industrial sectors. Its current share of commercial energy consumption is about 65%, com-

Table 16. Latin American coal reserves and resources

Country	*Geological resources* *Million tce*	*% of world*	*Technically and economcally recoverable reserves* *Million tce*	*% of world*	*Degree of exploration*
Brazil*	10 082	0.10	8098	1.27	L–M
Colombia*	8 318	0.08	443	0.07	L
Mexico*	5 448	0.05	875	0.14	M
Venezuela*	1 630	0.02	979	0.15	M
Peru*	1 122	0.01	105	0.02	M
Argentina*	384	—	290	0.05	L
Ecuador*	22	—	N/A	N/A	M
Bolivia	N/A	N/A	N/A	N/A	M
Costa Rica*	N/A	N/A	N/A	N/A	L
Chile*	4 585	0.05	162	0.03	M
Guatemala*	N/A	N/A	N/A	N/A	L
Honduras*	N/A	N/A	N/A	N/A	L
Panama*	N/A	N/A	N/A	N/A	L

Degree of exploration:
H = well established reserve estimates, well defined geology within each coal-bearing region, adequate coal-bearing data.
M = some exploration programmes have been documented.
L = few exploratory programmes with very little specific resultant data.
— negligible, less than 0.01%.
N/A = not available.
* coastal state.
Source: Coal Development Potential and Prospects in the Developing Countries. World Bank, October 1979.

pared to 31% for crude oil. The major use of coal is in thermal power generation which represents about 60% of a total power generation of 31 000 MW.

Since 1974, the Government of India has imposed restrictions on oil for power generation, and thereby promoted greater use of coal. In addition, the sixth Five-Year Plan calls for such coal-fired plant capacity to be increased by nearly 20 000 MW. The use of coal in power generation, consequently, has been projected by the Government of India to grow from 29.8 million tons in 1977–78 to 100.1 million tons in 1992–93 and to 158.7 million tons in the year 2000.[6]

Table 17. Asian coal reserves and resources

Country	Geological resources Million tce	Geological resources % of world	Technically and economically recoverable reserves Million tce	Technically and economically recoverable reserves % of world	Degree of exploration
India*	56 799	0.56	33 700	5.30	M–H
Indonesia*	3 723	0.04	1 430	0.22	L
Turkey*	3 268	0.03	793	0.12	M
Vietnam*	3 000	0.03	N/A	N/A	M
Bangladesh*	1 649	0.02	519	0.08	
Pakistan*	1 375	0.01	N/A	N/A	L–M
Korea, R.*	921	0.01	386	0.06	M
Taiwan*	680	—	N/A	N/A	M
Japan*	8 641	0.09	1 006	0.16	H
Iran*	385	—	193	0.03	M
Philippines*	87	—	N/A	N/A	L–M
Afghanistan	85	—	N/A	N/A	L
Thailand*	78	—	N/A	N/A	M
Brunei	1	—	N/A	N/A	L
Cambodia	N/A	N/A	N/A	N/A	L
China, P.R.	1 438 045	14.20	98 883	15.54	M
Korea, D.P.R.	2 000	0.02	480	0.08	M
Burma	286	—	N/A	N/A	M

Degree of exploration:
H = well established reserve estimates, well defined geology within each coal-bearing region, adequate coal-bearing data.
M = some exploration programmes have been documented.
L = few exploratory programmes with very little specific resultant data.
— = negligible, less than 0.01%.
N/A = not available.
* coastal state.
Source: Coal Development Potential and Prospects in the Developing Countries. World Bank, October 1979.

Indonesia has a potential similar to that of India to increase its use of coal. However, the gains foreseen are perceived to result more from substituting coal for oil in domestic electricity production so that oil exports can be expanded. While recoverable coal reserves amount to some 1430 million metric tonnes of coal equivalent (mtce), the greater portion of these reserves (some 1350 million mtce) are classified as brown coal.[7]

Table 18. African coal reserves and resources

Country	Geological resources Million tce	% of world	Technically and economically recoverable reserves Million tce	% of world	Degree of exploration
Botswana	100 000	0.99	3 500	0.55	L–M
South Africa*	57 566	0.57	26 903	4.23	M
Zimbabwe	7 130	0.07	755	0.12	M
Swaziland	5 000	0.05	1 820	0.29	M
Angola*	500	—	N/A	N/A	L
Cameroon*	500	—	N/A	N/A	L
Mozambique*	400	—	80	0.01	
Zambia	228	—	5	—	L–M
Tanzania*	360	—	N/A	N/A	M
Nigeria*	180	—	90	0.01	M
Madagascar*	92	—	N/A	N/A	M
Morocco*	96	—	N/A	N/A	M
Egypt*	80	—	N/A	N/A	M
Zaire*	73	—	N/A	N/A	L
Algeria*	20	—	N/A	N/A	L
Malawi	14	—	N/A	N/A	M
Benin*	N/A	—	N/A	N/A	L
Burundi*	N/A	—	N/A	N/A	L
Ethiopia*	N/A	—	N/A	N/A	L
Niger*	N/A	—	N/A	N/A	L
Sierra Leone*	N/A	—	N/A	N/A	L
Somalia*	N/A	—	N/A	N/A	L
Tunisia*	N/A	—	N/A	N/A	L
TOTAL AFRICA	172 239	1.71	33 153	5.36	

Degree of exploration:

H = well established reserve estimates, well defined geology within each coal-bearing region, adequate coal-bearing data.
M = some exploration programmes have been documented.
L = few exploratory programmes with very little specific resultant data.
— = negligible, less than 0.01%.
N/A = not available.
* coastal state.

Source: Coal Development Potential and Prospects in the Developing Countries. World Bank, October 1979.

Colombia is one of the few developing countries whose substantial coal resource levels and their geographic proximity suggest that it can greatly increase its coal consumption and production as well as its export. Since the exploration for coal in Colombia has only recently expanded, its reserve figure will continue to rise for some time. The majority of the coal deposits are in the eastern Andes mountains with the most important export-oriented deposit at Cerrejon in the Department of Guajira. This deposit is estimated to contain 1800 million tons of thermal coal reserves, all of which are amenable to exploitation through surface mining methods.

Although coal has been mined in Colombia for many years, the industry operates on a small scale. Production reached only 5.8 million tonnes in 1980 and most of this was consumed domestically in residential (thermal) and metallurgical (steel) uses. Since 1975, the government has taken a greater interest in coal's potential as an energy resource. According to the Plan de Integracion Nacional (1982), thermal coal consumption in electric power generation was to reach 4.0 million tonnes in 1985 and may reach 6.5 million tonnes in 1990.[8]

These plans, however, are modest compared to the foreseen potential of sizeable coal exports and their accompanying increase in export earnings. The government has thus begun a programme of joint ventures employing foreign capital and technology. According to Schneiderman (1981) and Zorn (1982), the most important of these is the El Cerrejon deposit presently being developed jointly by the International Colombia Resources Corporation (Intercor), a subsidiary of Exxon Corporation and Carbones de Colombia S.A. (Carbocol), an agency of the Colombian government.[9–11]

SUGGESTED POLICIES AND ACTION

Based on the above observations, some policies and actions would be required by the potential host developing coal endowed country to attract a nodule processing plant. Some of the most important include:

(1) Efforts at acquiring technical, economic and institutional information on how coal reserves, markets and technologies have to be initiated.

(2) Coal-based projects need to be incorporated into comprehensive regional and national development programmes with consideration given to attaining the efficient use of infrastructure between relevant sectors. This could involve the use of coal-fired plants to supply heat for other industrial concerns and communities.

(3) Technologically innovative approaches for coal handling, coal burning, plant siting and electricity transportation need to be pursued. Slurry systems, power barge siting systems and direct current transmission need to be examined, taking into account available capital, available information and the higher economic growth rates expected in developing countries.

(4) Advanced technologies and management for investigation, exploration and extraction of reserves should be introduced through effective co-operation with the industrial countries.

(5) Environmental regulations should be carefully designed to take account of cost-benefits while giving due consideration to the geographical, meteorological, ecological, industrial and social conditions of the region or country.

(6) Vigorous institutional arrangements for coal development should be implemented through effective incentives and streamlined government procedures. Development should be encouraged and systems established for joint venture participation or mine leasing.

SUMMARY

In the industrialized countries, policies and conditions with respect to coal and new coal development projects vary. Few problems are foreseen with domestic coal supplies and their use in energy supply in Australia and Canada, whereas some problems are foreseen in the United States (utility financing and air emission controls), Germany (inadequate resources), Netherlands (sensitivity to degradation of environmental quality and lack of space for coal handling and waste disposal) and Japan (environmental siting and waste disposal problems).

With regard to the benefits to be derived by developing countries through the development of coal deposits for power generation, it is generally recognized that the potential benefits could be wide-ranging. In addition to employment benefits, the major impact will be felt on the balance of payments. Coal development in developing countries for domestic use will reduce oil consumption, thus saving foreign exchange for net oil-importing countries, or releasing additional oil for exports in the case of oil-exporting developing countries. Tables 16, 17, and 18 summarize the available information on developing country coal reserves and resources.

Again, electric power is generated in thermal plants burning oil, gas or coal, and from hydro, geothermal and nuclear power plants. Which of these sources is selected depends on its availability and of the relative cost of the electricity generated. The cost of electricity varies not only with the fuel but with the type of technology used to generate electricity, the type of load, pollution control technology requirements and load factors.

In highlighting the resources and reserves of coal in developing countries, the objective is not to suggest that other sources are unacceptable but to outline some of the steps that would need to be undertaken by countries interested in developing a significant backward linkage to a nodule processing plant through utilizing domestic energy resources that could meet other development objectives.

Other potential backward linkages include the production of limestone and dolomite (which find other uses in, *inter alia*, agriculture, chemicals manufacture, etc.) and the development or expansion of other non-energy inputs into the processing plant assuming either the raw materials base is available or the service industry is in operation.

NOTES

[1] For the purpose of qualitative comparison, raw coal tonnages may be converted into tons of coal equivalent (tce), corresponding to the calorific value of one metric ton of hard coal, at 7000 kcal/kg. In addition, certain types of coal vary from country to country in terms of their tce value. Coal may also be compared with oil in terms of energy value and the following table contains some conversion factors that are useful in such a comparison.

Standard energy conversions for coal and oil

From/to	*tce*	*toe*	*boe*	*bdoe*
tce	1.00	0.65	4.80	0.013
toe	1.50	1.00	7.40	0.200
boe	0.21	0.13	1.00	0.003
bdoe	76.00	50.00	365.00	1.000

tce = tons of coal equivalent
toe = tons of oil equivalent
boe = barrels of oil equivalent
bdoe = barrels per day of oil equivalent

Source: *Energy: Global Prospects 1985 to 2000*, Workshop on Alternative Energy Strategies (WAES), 1977.

[2] Griffin, J. M. and Gregory, P. R. 1976. "An intercountry translog model of energy substitution responses". *Amer. Econ. Rev.*, Vol. 66, pp. 845–857.
[3] Pindyck, R. S. 1979. *The Structure of World Energy Demand.* MIT Press, Cambridge and London.
[4] OECD, 1978. *Steam Coal Prospects to 2000.* International Energy Agency, Paris.
[5] World Bank, 1979. *Coal Development Potential and Prospects in the Developing Countries.* World Bank, Washington, D.C.
[6] Government of India, 1980. Working Paper on Energy Policy. Controller of Publications, Delhi, Tabs. 11.4 and 11.5(b).
[7] Labys, W. C., Pachuri, R. K. and Blitzer, C. L. 1983. "Coal: a development challenge". *Natural Resources Forum*, Vol. 7, No. 4, United Nations, pp. 299–310.
[8] Ministerio de Minas y Energéa, 1981. Departamento Nacional de Planeacion, Republica de Colombia, Bogotà.
[9] Schneiderman, S. J. 1981. "El Cerrejon: Colombia's big new coal project", *World Coal* (September/October 1981), pp. 58–62.

[10] Zorn, S. A. 1982. "Recent coal and uranium investment agreements in the Third World", *Natural Resources Forum*, Vol. 6, No. 4, United Nations, pp. 345–358.
[11] Labys, W. C., Pachuri, R. K. and Blitzer, C. D., 1983, op. cit.

REFERENCES

American Society for Testing and Materials. 1982. *Standard Test Method for Shake Extraction of Solid Waste with Water*. D3987 in 1982 Annual Book of ASTM Standards: Part 31, Water. Philadelphia, PA.

Bosson, R. and Varon, B. 1977. *The Mining Industry and the Developing Countries*. Oxford University Press for the World Bank.

Callies, D. L. and Johnson, C. J. 1985. "A survey of land and environmental laws and business/economic factors applicable to the siting of a manganese nodules processing plant in the Pacific basin/rim: Australia, Canada, Colombia, Ecuador, Fiji and the Philippines." University of Hawaii Sea Grant College Program, Grant No. NA81AA-D-00070 Sea Grant Report UN1H1-SEAGRANT-CR-85-02, April 1985.

Dames and Moore and EIC Corporation. 1977. *Description of Manganese Nodules Processing Activities for Environment Studies*. Vols. I, II and III. United States Department of Commerce—NOAA, Office of Marine Minerals, Rockville MD. NTISPB274912.

Deep Seabed Mining—Final Programmatic Environmental Impact Statement. 1981. Vol. I. US Department of Commerce, National Oceanic and Atmospheric Administration, Office of Ocean Minerals and Energy, September 1981.

Flipse, J. E. 1983. "Deep Ocean Mining Economics" in Offshore Technology Conference, 2–5 May 1983 OTC 4491.

Flipse, J. E., Andrews, B. V. and Brown, F. C. 1983. "The economic viability of a four-metal deep ocean mining venture." Washington, D.C., National Oceanic and Atmospheric Administration.

Griffin, J. M. and Gregory, P. R. 1976. "An intercounty translog model of energy substitution responses", *Amer. Econ. Rev.*, Vol. 66, pp. 845–857.

Groo, E. S. 1971. "Choosing foreign locations: one company's experience", *Colombia Journal of World Business*, September–October 1971.

Grosse, R. and Stack, J. 1984. "Non economic risk evaluation in multinational banks", *Management International Review*, Vol. 24, No. 1.

Hawaii Department of Planning and Economic Development, 1981. *Feasibility and potential impact of manganese nodule processing in the Puna and Kohala districts of Hawaii*. Prepared by the Department of Planning and Economic Development, State of Hawaii and the US Office of Ocean Minerals and

Energy, Honolulu: Department of Planning and Economic Development, State of Hawaii.

Haynes, B. W., Barron, D. C., Kramer, G. W., Maeda, R. and Magyar, M. J. 1985. Laboratory processing and characterization of waste materials from manganese nodules. US Bureau of Mines, Report of Investigations RI 8938, 16 pp.

Herring, R. J. 1983. *Managing International Risk.* Cambridge University Press, New York.

Hillman, C. T., 1983. Manganese nodules resources of three areas in the Northeast Pacific Ocean with proposed mining-beneficiation systems and costs. Washington, D.C., US Department of the Interior, Bureau of Mines Information Circular 8933.

IMCO/FAO/UNESCO/WMO/WHO/IAEA/UN. Joint Group of Experts on the Scientific Aspects of Marine Pollution (GESAMP) 1977. *Scientific Aspects of Pollution Arising from the Exploration and Exploitation of the Sea-Bed.* GESAMP Reports and Studies, No. 7.

International Atomic Energy Agency, 1978. *Steam Coal Prospects to 2000.*

Jaenicke, G., Scanze, E. and Hauser, W. 1981. *A Joint Venture Agreement for Seabed Mining.* Frankfurt am Main: Metzner; Deventer, The Netherlands: Kluwer (Studies in transnational corporations).

Johnson, C. J. and Otto, J. M. 1986. "Manganese nodule project economics: Factors relating to the Pacific region", *Resource Policy*, Vol. 12, No. 1, March 1986.

Labys, W. C., Pachuri, R. K. and Blitzer, C. L. 1983. "Coal: a development challenge", *Natural Resources Forum.* Vol. 7, No. 4, United Nations, pp. 299–310.

Nyhart, J. D., Triantafyllon, M. S., Avaback, J., Blick, A., Sklar, B., Gillia, M., Kirkpatrick, D., Muggeridge, J., Nakagawa, R., Newman, J. and Will, A. 1983. *A Pioneer Deep Ocean Mining Venture.* MIT Sea Grant College Program, MITSG83-14, 255 pp.

Pasho, D. W. and King, D. E. C. 1980(a). "Processing sector cost estimates for the Kennecott joint venture manganese nodule processing facility." Mineral Policy Sector Internal report MRI 82/5. Canadian Department of Energy, Mines and Resources, August 1980.

Pasho, D. W. and King, D. E. C. 1980(b). "Processing sector cost estimates for the Ocean Management Inc. manganese nodules processing facility." Mineral Policy Sector Internal report MRI 82/7. Canadian Department of Energy, Mines and Resources, August 1980.

Pindyck, R. S., 1979. *The Structure of World Energy Demand.* MIT Press, Cambridge and London.

Preparatory Commission for the International Sea-Bed Authority and for the International Tribunal for the Law of the Sea, 1986. The Enterprise: Economic viability of deep sea-bed mining of polymetallic nodules.

(Document LOS/PCN/SCN.2/WP.10, 14 January 1986, prepared by Ingham, P.D.).

Schneiderman, S. J. 1981. "El Cerrejon: Colombia's big new coal project", World Coal (September /October 1981), pp. 58–62.

United Nations, 1983. *The Law of the Sea.* Official Text of the United Nations Convention on the Law of the Sea with Annexes and Index, United Nations, New York, E.83. V.5.

United Nations, 1984. *Analysis of Exploration and Mining Technology for Manganese Nodules.* Seabed Mineral Series, Vol. 2. Graham and Trotman (London) 140 pp.

United Nations, 1986. *Analysis of Processing Technology for Manganese Nodules.* Seabed Mineral Series, Vol. 3. Graham and Trotman (London) 79 pp.

US Department of Commerce, 1979. "Gulf Coast Manganese Nodules Processing Plant Location Criteria."

US Environmental Protection Agency/Corps of Engineers. 1977. *Ecological Evaluation of Proposed Discharge of Dredged Material into Ocean Waters.* Environmental Effects Laboratory, Waterways Experiment Station, US Army Corps of Engineers, Vicksburg, MS. July 1977, 103 pp.

United States Federal Register. Parts I–IX, Environmental Protection Agency: Hazardous Waste and Consolidated Permit Regulations. Vol. 115, No. 98, May 19, 1980, Book 2 pp 33063–33285, 110-CFR, Parts 260–265.

Workshop on Alternative Energy Strategies (WAES), 1977. *Energy: Global Prospects 1985 to 2000.*

World Bank, 1979. *Coal Development Potential and Prospects in the Developing Countries.* World Bank, Washington, D.C.

World Bank, 1983. *Private Lending to Developing Countries.* Washington, D.C. World Bank Staff Working Paper, No. 482.

World Port Index, 1984. Pub. 150, Ninth Edition, Published by the Defence Mapping Agency, Hydrographic/Topographic Center.

Zorn, S. A. 1982. "Recent coal and uranium investment agreements in the Third World", *Natural Resources Forum,* Vol. 6, No. 4, United Nations, pp. 345–358.

List of Figures

List of Tables

INDEX

MIX
Papier aus verantwortungsvollen Quellen
Paper from responsible sources
FSC® C105338

If you have any concerns about our products,
you can contact us on
ProductSafety@springernature.com

In case Publisher is established outside the EU,
the EU authorized representative is:
Springer Nature Customer Service Center GmbH
Europaplatz 3, 69115 Heidelberg, Germany

Printed by Libri Plureos GmbH
in Hamburg, Germany